IF GOD EXISTS

IF GOD EXISTS

The Theory of Omnideism:
An Atheistic Argument for the Existence of God

M. E. Stearn

RESOURCE *Publications* · Eugene, Oregon

Resource Publications
An Imprint of Wipf and Stock Publishers
199 W. 8th Ave., Suite 3
Eugene, OR 97401

www.wipfandstock.com

PAPERBACK ISBN: 978-1-5326-1437-8
HARDCOVER ISBN: 978-1-5326-1439-2
EBOOK ISBN: 978-1-5326-1438-5

Manufactured in the U.S.A.   MAY 10, 2017

CONTENTS

Preface

THE ZYGOTE OF OMNIDEISM was formed at the age of seven, when my mother disabused me of the notion that Mary was really a virgin. From that day I obstreperously and publicly questioned the doctrine, dogma and authority of the Church, which tended to cause no little consternation to my Irish Roman Catholic family. A life-long theological and philosophical examination, inevitably, led to the most basic questions as to the explanations and arguments for God's existence or non-existence presented by both theists and atheists. None of the arguments, which the opposing sides propose to support their case, have provided an answer that, for me, is logically, philosophically, theologically or scientifically satisfying. It was this vacuum that compelled me to syllogistically conclude that: "If God actually exists, then the proof of God's existence must equally satisfy scientific and theological concerns." This simple statement has resulted in the audacious and paradoxical theory of Omnideism.

1

The Case Against Faith

THE APOSTLES' CREED, WHICH was once necessary for one to memorize in order to be confirmed in the Catholic church, states:

> "I believe in God,
> the Father almighty,
> Creator of heaven and earth
> And in Jesus Christ
> His only son, our Lord . . .
> Born of the Virgin Mary . . . "

Faith is the belief that something is true without requiring any facts to support that belief. Therefore "faith", by definition, questions the actual existence of that which one professes to believe because there is, and can be, nothing to support the belief.

If one believes and God doesn't exist then belief is irrelevant.

If one believes and God exists then belief is irrelevant because the fact of the existence renders the belief moot.

If one doesn't believe and God doesn't exist belief is irrelevant.

If one doesn't believe and God exists belief is irrelevant because the fact renders the belief moot.

Nobody has faith that $1 + 1 = 2$ because it is an absolute fact. The belief in an irrefutable fact is irrelevant to one's faith. Whether one has faith or not in a certain belief does not alter a fact. If nobody in the world believed 1 + 1 = 2 this would not change the fact. Just as believing that the Sun revolved around the Earth did not change the fact. People don't profess faith that the earth is round, or that the heart pumps blood through the body, or in atomic particles. They are all facts. A fact exists independent and with no relationship to one's faith. Faith and fact cannot logically co-exist regarding the same belief. Faith can only exist absent facts. Faith cannot exist where facts exist because a fact extinguishes the belief upon which faith is based. Faith requires the absence of fact.

God either exists or He doesn't exist. For theists the idea or concept of God is necessary. The actual existence of God is irrelevant. But if God exists then the idea or concept of God is irrelevant. Omnideistically only the actual existence of God is necessary.

Theists need to believe there is a creator. Theists need to believe that there is a higher power to whom they are beholden. Theists need to believe there is a God to whom they can speak. Theists need to believe. Belief is the core of all religions because belief is proof of one's faith and loyalty. Belief is proof of one's faith absent evidence of fact. Each religion has its own peculiar set of beliefs based on a particular concept or idea of God. The quantity of gods equals the quantity of concepts.

The vast majority of people have faith in the existence of some kind of supreme spiritual being which they also insist is a fact. Yet if God is accepted as fact, or even a provable theory, then there would be no need or purpose to profess faith in the belief. No religion organized or otherwise, puts any significant emphasis on the actual existence of God other than to say His existence is evident from what we experience in daily life. Religion is about belief and faith. Proof of one's faith is the foundation of every religion. Acts of faith are demonstrations that one truly believes in a Supreme Being for which there are no supporting facts which prove the existence of the Supreme Being. Acts of faith would be purposeless if the Supreme Being actually existed.

Conversely faith as used by humans provides the freedom for any action. The faith of the 9-11 jihadists, the faith of the thirteenth and fourteenth century crusaders, the faith of Christians, Jews, and Muslims has given them all the freedom to use force against each other as well as countless intra-sectarian affairs.

Yet faith, because it can only exist absent facts, calls into question the actual existence of the thing that the faithful call on to support their actions. The most fundamental Christians use faith to deny the age of the universe, the evolution of humans, and the process of climate change because facts are anathema to the faithful. Faith is both a barricade and a weapon to those who hide behind it and wield it indiscriminately in their attacks.

Whether it is a Muslim, Jew, Christian, Hindu or animist speaking each will say something to the effect of "That is not the God that I believe in" when asked about the God which is the object of faith in another religion. Taken literally, this is confirmation that there is a different God for each theology. But every theist will quickly stress there is only one God (or Gods), theirs. What should be originally stated is, "That is not the idea or concept of God that I believe in."

Faith and fact are irreconcilable. Facts negate the necessity for faith. The faithful claim that God must exist because one can see the results of God's existence, i.e., the stars, Earth, life, beauty etc. Nobody can see the speed of light or sub-atomic particles yet everybody can see the results of their existence. Scientists never suggest that he or she has faith in the existence in the speed of light or in sub-atomic particles. Yet irrefutable proof that God actually exists would utterly undermine the necessity for faith. So theologically/philosophically the question is: Can God exist absent faith?

A true believer will always resort to fending off any attack, which questions the existence of God with, "I don't care what the facts show (or "what you say", "what your rationale is" etc.), because this is what I believe, I have faith". The passion displayed in defending a belief unsupported by fact, is far beyond anything that might be shown in defending a fact, because there is no need to defend a fact. Children are taught early to fight for what they believe. Children are never taught to fight for a fact. At their most fundamental level wars are not fought in order to defend a fact they are fought to defend a belief. Belief in the moral superiority of one's politics, religion, and ethnicity are the rationalities for war. Fighting over land is done because one group or tribe, or religious sect, or ethnicity believes it is superior to the other and therefore deserves the land. We have only fought wars to defend positions for which there are no supporting facts. Even World War One, the Hapsburg Family Feud, was instigated upon the false belief that the Austria-Hungarian Empire had the right to subjugate a minority Serbian population because of its moral superiority. The eventual involvement of Russia, France and England as the result of treaties, may be the exception that proverbially proves the rule.

The depth or sincerity of faith in the existence of God cannot change the fact of God's existence or non-existence. Conversely, the actual existence (the fact) of God, by definition, extinguishes faith based upon belief in God. Belief in God's existence or belief in God's non-existence is not an issue. God either exists or not. But if God exists then he can only exist consistent with either a theory supported by some set of facts which prove His existence or because of an absence of facts denying His existence and satisfactorily explain the existence of the Cosmos.

The cornerstone of every religion is the absolute unquestioning faith of its followers in that religion's concept of God. Every religious sect is founded on the faith, rather than the actual existence of God (or Gods). Faith in God's existence is based upon the need to explain the existence of, and the purpose for, the existence of the Cosmos. Faith by its very nature puts into question the existence of God. If faith is taken out of the equation how does that effect God's existence? Faith in Mary's virgin pregnancy was the basis for a war within the Christian faith for centuries. The Islamic world is in an intra-sectarian war today because of faith.

Whether or not God actually exists, the Cosmos does exist. Billions of galaxies, each with billions of stars exist. Energy and matter in its infinite forms exist. Faith in God's existence does not change the actual existence of the Cosmos. Faith exists only to satisfy the questions to which humans have so far been unable to answer with a logical or scientific explanation. Although science cannot directly provide an answer to the purpose of our existence, science and logic can provide us with a roadmap for our own existence. If we can logically or scientifically explain how we have come to exist we can pursue the reason for our existence based on the underlying facts. Faith will only be relevant to how we believe we should act consistent with the fact of God's existence.

The argument that faith is necessary for people to know how to act in order to please God is misplaced. Faith in God does not satisfactorily explain why humans, generally, don't kill, rob, rape, and pillage but rather that such acts are contrary to supporting the continuation of the species. If faith in God's existence was the primary reason for people positively interacting then atheists, historically, would be the world's leading perpetrators of war and terrorism instead of the most religiously "faithful" whether it was Jupiter, Zeus, Thor, Yahweh or an animistic god. Combatants, on both sides of any type of war, always have faith that their cause is the one supported by God. Faith is used as a license to perpetrate any act of violence or torture or abuse on another human being. And it is not just war that faith is used as a justification for abominable acts. The treatment of women throughout most of Africa, the Mid East and Southeast Asia including India is based upon theistic faith.

Faith in facts being false may require more faith than belief that a higher deity created the universe over 7 days sans dinosaurs. Accepting the speed of light as a fact and seeing stars in the sky that are 10,000 light years away and believing that the universe is only 6,000 years old requires an extraordinary type of faith.

A literal translation of the Bible infers a magical sort of God acting as a Houdini or David Copperfield. Taking mud and creating man and then taking a rib from the man and creating woman is nothing but abra-cadabra time. Creationists, by definition deny the laws of physics by denying that the galaxies that exist billions of light years away from earth actually don't exist because the universe is only 6000 years old. So Creationists must have faith that the laws of physics are false. Well not all the laws. Not even the most ardent Creationist will question the fusion that produces heat and light from the sun. Nor will they question that it takes 8 minutes for that light to reach the Earth. Nor do Creationists question the existence of any of the elements or the sub-atomic particles. Nor do Creationists question the physics that explain the orbits of the planets around the sun. In fact Creationists will argue that it is God who created the laws of physics and chemistry that allow the sun to shine and the elements to be formed.

What Creationists can't explain is why would the same God who created the laws of and chemistry have to resort to magic to create humans or the Earth or anything else when He had already created the mechanism? Why wouldn't God be confident enough in the natural laws of science, which He created to allow those laws to work as the scientific community theorizes? Why would the God of Genesis, having already created all the animals, females and males, forget the very next day that in order to procreate Adam needed Eve? The fundamentalists' literal God is a blasphemous interpretation of the Intelligent Designer who created the very laws that the fundamentalists disavow. Why do fundamentalists insist that the belief in laws created by God are of lesser value than belief in Creationism? Belief in creationism requires no facts, only faith. Belief in the laws of physics is based entirely on facts. Absent faith creationism cannot exist.

Although the form of creation varies from sect to sect, and even from person to person within each sect, the concept of God, as the creator in control of the beginning of the formation of the universe, is constant. Ironically, it is the creationists who, by definition of faith, put into question the actual existence of God as opposed to an atheist who puts into question the non-existence of God. Atheists of course are in the stronger logical position by stating that "absent evidence of His existence" they cannot believe in God.

As a reasonable and rational intelligent creator, God needed only to have created all the atomic or sub-atomic particles which would eventually form a black hole which, upon reaching the critical point, the Hawking singularity, resulting in the Big Bang. It certainly would fall within the perception of an omniscient God that He could initiate such a creation with full knowledge of the ultimate results caused by the laws of physics, chemistry and biology, which He created.

Some fundamentalists latch on to things like intelligent design in order to help bridge the gap between creationists and evolutionists because the creationists claim that the development of the nervous system is too complicated to be explained by evolution. This of course implies that evolution, consistent with the laws of biology, which He created, is too difficult a concept for God, as the omniscient and omnipotent creator, to have utilized because He wasn't smart enough to understand the scientific laws of nature which resulted in the evolution of all species.

Creationism is more about faith that facts do not exist rather than faith in the concept of a creator God, whom they believe exists. A creationist must believe that the Big Bang didn't occur, that the stars and galaxies we see millions of light years away don't exist (or alternatively the speed of light is an optical illusion and hoax), that carbon dating, and other radiation dating through various elements, is a hoax, that the dinosaurs, wooly mammoths, and saber-tooth tigers that existed were overlooked by Biblical writers (not to mention Noah). A creationist's greatest faith is the faith of denial and deliberate ignorance.

A rational version of Genesis which includes God the creator could be re-written thus:

> "In the beginning God created matter and energy, which He compressed into a tiny ball so small that it finally exploded into the nothingness creating all the elements of the universe according to His laws of nature. The matter and energy eventually formed the Heaven and the Earth and an infinite number of other bodies speeding across and filling the empty universe including the sun and the stars which shone upon the Earth. Then as a result of God's natural laws the matter and energy coalesced into simple forms of life on Earth as it rotated and revolved around the star God called the Sun. Some of the matter and energy which didn't become part of the Earth developed single cell forms of life which begat according to

His laws, multiple cell forms which begat more complex forms of life including plants and animals until such animal forms could move about in the water and about the Earth. These animal life forms begat more complex life forms, male and female, evolving into a great variety of species some of which developed greater intelligence and understanding of their own beings one generation to the next until there evolved a multitude of intelligent animals. Some of these animals begat other animals and over the generations some of these animals began to reason and understand the Lord and these animal forms God called Human Beings.

And God's natural laws caused the rain and the snow to fall and the hail and caused the mountains to erupt and the earth to shake and the wind to blow and the sun an stars to shine and the Earth and other planets to orbit and rotate."

The problem with this "rational version" is found within the very first phrase, "God created". Everything in the Cosmos that exists, exists as some form of matter and energy. Therefore, if God is the creator then He must exist. And if God exists then it follows that God exists in some form of matter and/or energy. The only other possibility is that God exists as a vacuum, absent of any matter and/or energy. By definition nothing exists in a vacuum therefore God can only exist as some form of matter and/or energy. God is the creator of the Cosmos by theistic definition. Therefore God either exists in some form of matter and/or energy or God does not exist. But if God, as the creator, created matter and energy how could God create the thing of which he exists? Therefore, if God exists as some form of matter and energy then God could not have created matter and energy since it could not have existed before matter and energy were created. Therefore God could not have existed before He created the stuff of which He is made.

God *is nebulous*. Theological arguments that postulate that God exists, but doesn't necessarily exist in some form of matter and/or energy, must rely on faith, based on the unsupported belief that something, other than matter and energy, exists in the cosmos.

That faith based argument is founded completely on the denial of scientifically supported facts, in order to sustain a belief that undermines the object of the belief, God. The general definition of the verb "to exist" in English language dictionaries is that the thing that exists, is an actual being or thing in either material or spiritual form (Merriam-Webster). Even if God is defined as a spirit then He still exists as some form of energy, even if that form of energy is nothing more than Higgs Bosun (the God particle). There is no evidence of any kind that anything exists in the Cosmos that does not consist of either matter and/or energy unless one argues that an idea or concept exists as an actual thing.

An idea does not consist of photons, electrons, quarks or any other sub-atomic particles. An idea or concept does not exist as any form of matter or energy but the result of a process that involves matter and energy. If God is an idea or concept then God does not exist. If God's nebulous state existed only as a concept then there can be as many God's as there are ideas. If God exists as an idea then Santa, the Tooth Fairy, and Huckleberry Finn also exist.

The argument that God is formless is also consistent with the way some describe the cosmos prior to the formation of the singularity which preceded the Big Bang. But even a formless God must exist in some form of energy/matter even if that energy is sub-atomic particles randomly moving in space.

God either exists or God doesn't exist. God is either a fact or a concept. Faith, or the lack thereof, does not change a fact. If God is just a concept then, any expression of faith, is legitimate. Paradoxically, every religion requires that only their particular concept of God is exercised in order to defend their faith. Every religion believes their idea of God is the only legitimate perception of the deity. Every monotheistic religion believes that every other religion's concept of God, and what they believe that God wants, is blasphemous. Everyone's faith requires that all the other faiths' concepts of God are, at the very least, partially wrong. Faith can only be used to defend or espouse a concept. Faith cannot be used to defend or espouse a fact.

If God is a fact then only one concept, perfectly congruent to the actual existence of God, is legitimate. Therefore, faith can only exist if God is a concept. Faith cannot exist if God is a fact. Even if God's existence is characterized as spiritual, and neither matter nor energy, then, again, God's existence is relegated to a concept rather than an actual presence. Can one have faith in God as just a concept? Yes, but only if one does not believe in God as the Creator. God as the creator defines God as actually existing, a being constituted in some form of matter and energy. God as a concept cannot be the Creator. If God is the Creator He must actually exist and therefore theistic faith is irrelevant.

2

The Case Against God as
the Creator

Definition of God—Merriam Webster defines God as the one who
created and rules the universe.

The Oxford-English and Random House Dictionary defines God as
the creator and ruler of the universe.

ATHEISTS AND FUNDAMENTALISTS DO have a unique point of agree-
ment. Evolution proves everything. If you are an atheist evolution
proves there is no God and if you are a Fundamentalist the existence of God proves there is no evolution. Atheists theorize that if
evolution is true then God could not possibly be involved because
the laws of nature supplant the necessity of God. Fundamentalists
cannot seem to grasp the idea that evolution is as miraculous if
not more so than the abra-cadabra of the biblical creation. The 'intelligent designers" cite the complexity of evolution as disproving
evolution, inferring that God is not nearly intelligent enough to
understand and create through biological evolution. Atheists argue
that evolution is proof of God's non-existence presumably because

it contradicts the biblical interpretation. The Fundamentalists should be arguing the converse, that the complexity of evolution proves the existence of a supreme being.

Every monotheistic religion that believes in God defines God as the creator of the universe, of everything that exists. This definition postulates that God must be infinite since, by definition, God cannot have been created. The first lesson I remember from the Baltimore Catechism was "God always was and always will be". The eternal nature of God allows God to act as the creator but does not define God as a creator, because although God is eternal it does not necessitate that God be a creator. Yet, conversely defining God as a creator requires God must be eternal, timeless, and infinite. God cannot be the creator if He is not eternal but He may be eternal although He is not the creator. Therefore, if God exists, the absolute essence of God is that He is eternal.

We must conclude two things: If God exists then God must exist as some form of matter and energy and; God must be eternal. Therefore, either God was created from existing matter and energy or God created matter and energy. God couldn't create matter and energy because that is the stuff from which He is formed, the stuff had to already exist. But God couldn't be created from existing matter and energy if God is eternal. And if God is not a form of matter and energy then God cannot exist. So God is either not the creator or God is not eternal. If God is not eternal then something created God and therefore God cannot be "the" Creator. God can only be the creator if He is eternal.

One of the alternatives to God, as the creator, is Hawking's theory of spontaneous combustion resulting in the Big Bang but that fails to explain what existed and created the singularity prior to the Big Bang. No physicist, including Hawking, can explain how the singularity was created. In fact Hawking's theory infers that Newton's first law didn't exist until the Big Bang which means that energy couldn't be created or destroyed until energy was created. This is as illogical as any version of Creationism. Instead of some supreme magician going "abra-cadabra there is the universe", the atheistic and agnostic physicists claim that energy just suddenly appeared, magic sans the magician.

The actual existence of God, as the creator, could be logically argued in the form of a negative proof since there has been no scientific theory or hypothesis offered to explain the formation of the singularity from the formlessness of the energy that must have existed under the First Law of Thermodynamics (a.k.a. the law of conservation of energy). Nor is there any hypothesis or theory as in what form energy existed prior to the formation of the singularity. The absence of a scientific explanation of any kind allows a Holmesian deduction that once all impossibilities are eliminated, that which remains, no matter how improbable, must be the truth.

The law of conservation states, unequivocally, that energy can be neither created nor destroyed and therefore the spontaneous appearance of all the energy that now occupies the Cosmos, in the form of the singularity, is impossible. The only argument is the law of conservation isn't true. But the evidence that supports the law of conservation has never been challenged with contrary evidence. Therefore, if energy cannot be created, according to the law of conservation, then the only possibility, no matter how improbable is that energy is infinite.

If God exists, then he exists as some form of matter and energy. Then must we conclude that matter and energy created God? If matter and energy created God then there is no reason for the existence of God the creator. If God could be created through the natural laws then the acts of God as a creator would be irrelevant. It may be argued that God was created from the formless sub-atomic particles and energy which existed, then, as the Creator, formed the cosmos from the remaining formless matter and energy which already existed at the time God Himself was formed. But this explanation then denies God's eternal existence and, thus, the essence of God. The creation of God the creator is impossible, if God exists.

Neither pure logic nor natural laws of physics allow for God as the ultimate creator of matter and energy. But if God is not the creator of matter and energy could God exists as a separate parallel form of matter and energy and "Our Father" Who created the laws which allowed the matter and energy to form the Cosmos? This creator God must possess an anthropomorphic consciousness that permitted Him to create the laws of physics, chemistry and biology. (It is very arguable that God only needed to create the laws of physics since without the formation of the elements the laws of biology and chemistry need not exist until after the Big Bang) Even that limited anthropomorphic quality that would make God the Creator (regardless of how limited an entity) who actually exists in some form of matter and energy.

But if God consists only of energy that energy cannot be even in the form of an electron since an electron is a particle that can be measured by mass. The same is true for a quark or positron. Is God a photon? A packet of energy? Or is God neither matter nor energy, but merely an idea or concept? Thoughts are not tangible. Nor are thoughts energy as a matter of physics. Thoughts, ideas, memories and concepts, however, are all the result of energy produced by matter, synapses, neurons, etc., contained in the brain. Thoughts cannot be substituted anywhere into the equation $E=mc(c)$. Is a thought eternal from the moment it is created? Once a thought is created can it be destroyed? A thought is not infinite since it must be created and therefore has a beginning. Is a thought created only when the brain creates it or must a thought be communicated in order to exist? If a thought is never communicated and the synapse or neuron, where the thought resides, dies, does the thought continue to exist? Does God exist as a concept, idea or thought?

The fundamental difficulty with an entity existing separately from the rest of all other matter and energy is that the formation of the extraneous matter and energy into a separate conscious entity, again, negates the infinite nature of God. If some pre-singularity of matter and energy could be formed into a conscious entity (God) then the formation of the singularity from the chaotic existence of sub-atomic particles and energy, (Higgs Bosun, quarks, photons, etc.) which resulted in the big bang would dictate that God was a creation of matter and energy contradicting His infinite nature. Absent an eternal nature, God's existence is meaningless because it no longer allows for any acceptable definition of God, because then God is no different from any other form of matter and energy. For God to exist as a separate entity to the rest of the Cosmos requires that God is created from matter and energy which again negates an infinite existence. God's infinite nature makes His existence as a separate entity in any form impossible.

Could God's existence be a parallel infinite plane to the existence of all other matter and energy and therefore God is in the position of creator, in the narrow sense, that He created the natural laws that govern the cosmos, and thus allowed for the formation of the singularity from the chaotic and formless sub-atomic particles (matter and energy), which existed prior to the big bang, as the conscious being who generated the big bang, again putting God in the position of an omnipotent and omniscient being, responsible for every single occurrence in the cosmos?

Can god the creator exist without being God the Father? Does God the Creator determine God the Father? Does God the Father determine God the Creator? Can God the Father exist without being God the Creator? Does God exist as father/mother, but not as a creator, no more responsible for our acts or the natural consequences on earth than our gene specific parents are responsible for our individual actions?

What is the rationale for God to exist if God is not the creator? Insisting that God must also be the Creator only acts to simplify God. Those who believe in a creator God inevitably believe in a very personal God, God the Father. One with Whom they can directly communicate. One Who is utterly involved in the every day activities of every individual. A personal God is anthropomorphic since personal communication with a supreme being, absent the essential human qualities, would be senseless. Traditionally, the Norse, Roman, Greek and Judeo/Christian, Islamic conception of God (or gods) resembles a physical human entity with a consciousness, an entity that possesses and responds with human-like thoughts and emotions. Does belief in a Creator God determine the belief in a personal God?

Does even the most limited version of God as a parallel infinite entity Who created the natural laws of the Cosmos, mean that this separate consciousness must, by definition, be omnipotent and omniscient? Could God exist, in the form of a non-involved, impassionate, impersonal entity Who did nothing more than create the laws that allowed the formation of the singularity and the resulting Big Bang?

Even this impersonal and remote type of creator God must be omniscient and omnipotent. It defies any form of logic that God could have created anything, either directly or indirectly, without knowing and understanding all the ramifications of that creation. Anything less than omniscience and omnipotence would redefine the entity of God as the Creator. Could an infinite being not be omniscient? Could the Creator of the Cosmos not be omnipotent? God as an infinite being, therefore, by definition, without restrictions of time, must be omniscient. A creator must, by definition, be Omnipotent as the only entity with the ability to create a comprehensive Cosmos. Anything less than omnipotence results in the possibility of some other entity possesses some power that escaped the Creator. Any such entity would mean that the Creator cannot be defined as God since another entity can control something in the Cosmos over which God has no control, and therefore the creator entity could not be God by definition. It can only be concluded that if God is the Creator in any form then God must be omniscient and omnipotent.

An omniscient/omnipotent God is either explicitly involved in every cosmic event by deliberately causing the event or He is implicitly involved because He is allowing the event to happen although He has full knowledge of the consequences. If God's only involvement was the creation of the natural laws that resulted in the Big Bang, His omniscience of all future events following that creation and therefore his intentional intervention, in even that single event in creation of the universe, means that He has deliberately not involved Himself in every other incident in the cosmos although he could have done so and knows the consequences of human pain and suffering as a result of not doing so. If God is involved in even one single aspect of any human being determines that He is personally responsible for every single event occurring in the Cosmos.

A tsunami is headed to shore. Thirty thousand people, who are in the way of the tsunami, all pray to God to intercede on their behalf. As a result of God's deliberate intervention one hundred fifty people survive. It is a miracle. As a result of God's deliberate failure to intervene twenty-nine thousand eight hundred fifty people die. God knew that the tsunami would kill everyone in its wake unless He saved them. God deliberately did not save 29,850 people from death nor did he save tens of thousands of more from injury and disease, nor hundreds of thousands who suffered from the loss of or injuries to loved ones. One hundred fifty survivors and their relatives and friends celebrate God's goodness. None of them consider that God was responsible for the tsunami to begin with.

In response to tsunamis and earthquakes and genocides, religious leaders encourage worshippers to thank God for His blessing in saving the survivors from death while simultaneously thanking Him for accepting those victims, who died and entered heaven. The survival of some of the victims may be called miraculous by believers even though the miracle just postponed their entry into heaven (or its sectarian equivalent) which is the ultimate goal of a theistic religion. Theists may argue that God did not deem the survivors ready for heaven and/or that we should not question the acts and motives of God, because God is perfect in his benevolence and/or He is unfathomable and/or humans have free will and/or it is necessary in order to justify good and/or human's deserve pain and suffering because of the original sin of Adam and Eve.

An unfathomable God is the safest answer to otherwise un-explainable death, pain and long-term suffering, because it allows God to act irrationally. The problem is that those who argue that the omniscient, omnipotent creator is unfathomable also argue that they "know" God loves us, they know God listens to our prayers, they know God is essentially good, and they know God has a plan. They know this unfathomable God because this is what God has revealed to them specifically.

All theistic religions proscribe to an anthropomorphic God who, at the very minimum, listens to, understands, responds to and cares about human beings as both a species and individually. The most quintessential response from God is a miracle. A miracle is always used by theists as absolute proof that God exists. Yet if God is an omnipotent creator then a miracle must be defined as an act which requires His intervention in order to reverse a previous act, which He caused. A miracle by that practical definition is either an apology or a correction, which means God had to make a mistake. If God is an omnipotent/omniscient Creator, then a miracle is no different than the heroic actions of the individual who saves the baby from the house that he set on fire.

One alternative to this explanation, if you are a theist, is to believe that God isn't in total control and some other entity may have caused the natural disaster. Another explanation would to believe God isn't in control of things on earth and things just happen but then you must also believe God isn't omniscient and didn't have any idea this or that disaster would happen. So either God isn't omniscient or God isn't omnipotent, and that natural disasters come as a surprise to God. But if we attribute even one miracle to God the Creator, one intentional intervention, then we must also attribute his lack of intervention as a deliberate act to allow pain, suffering and/or death to occur. If he had the power to intervene on behalf of anybody then his failure to use that same power must be measured on the same scale of His infinite love and compassion. The theists' counter-argument is that God is allowing those who die to enter heaven sooner and is withholding heaven as a result of his miraculous intervention.

For the faithful it is essential to believe that God's nature consists, at the very least, of the human qualities that, theologically and spiritually, allow the faithful to communicate with Him. Absent this belief, theists have no foundation on which to build their theological communities based upon the belief in God as the Creator. Not just a creator of the physical cosmos but the Creator and final judge of the cultural morals on which much of humanity bases the secular social contract.

A creator is theistically necessary because, without it, the existence of God would not hold the power to which the faithful ascribe to the supreme being, which is exactly the same perception small children have toward their mother and father. Children's relationship to God is founded upon the relationship to their parents and establishes the first anthropomorphic association we make with God and that perception never really changes for those who continue to believe in God the Creator. Every theist of faith believes that God has a consciousness that embraces human values and morals and, that based upon those values and morals God makes decisions and renders a final judgment that is the essential element to the purpose for human existence. Because the Creator renders the final judgment we must be eternally grateful and demonstrate that gratitude through various acts of acknowledgment.

All theists, all agnostics, all atheists use a phrase like, "Thank God". It is culturally ingrained. People all over the world are thanking God for some event or good fortune which has come their way. People all over the world believe that they are personally blessed by God because they possess a house, food, a job, health, a new car etc. What appears to be a simple act of gratitude by the faithful is fraught with inference.

Whether it is helping the little old lady across the street or the neighbor to shovel her sidewalk, or a giving few bucks to the homeless, the act is done deliberately without expectation of anything in return except "Thank you". The key word here is "deliberately": Acting with forethought and intent with knowledge of all the consequences. If somebody asks you to help them lift a box, or paint a fence a person responds knowing that their actions will make the requestor's life a bit easier. Usually there is no conflict of interest. Helping one person doesn't, in most circumstances, cause any harm to another, physically or emotionally. Yet every time we thank God we are inferring, that, for whatever thing for which we are giving "thanks", it was a result of God's deliberate and intentional undertaking to grant a wish, respond to a prayer or act of devotion in some other manner consistent with what we wanted. We are reacting to God the Creator as a conscious being, a being, who has carefully considered all the consequences of His specific action.

But thanking God for one's good fortune infers that God is responsible, not only for everyone's good fortune, but for everyone's misfortune. But unlike most human exchanges of gratitude, exchanges with God are fraught with conflicts of interest because of God's perceived position as the omnipotent/omniscient Creator.

An economic downturn, which results in millions of people suddenly being out of work, or more specifically looking for jobs, exemplifies the conundrum that the theistic faithful confront: that of seeking and receiving or not receiving and then responding to God's blessing or lack thereof.

Somewhere one hundred people provide resumes for one position and all of them pray to God, Allah, or Jehovah, yet only one of them will get the job. The job doesn't necessarily go to the individual who most needed it or to the one who prayed the most or hardest. Yet whoever gets the job will inevitably utter the phrase, "thank God", even the one atheist in the group. If God chose the winner why shouldn't the losers feel condemned because the Lord forsook their entreaty? Ninety-nine people are suddenly placed in a worse position than they were before because they expected to get the job, as a result of their entreaties to their Creator, and now do not have a paycheck because their prayers were ignored and they cannot pay for the mortgage, food or clothes. And what if the only atheist in the bunch who never uttered a prayer or a thought of God got the job, what can the faithful conclude?

The popular phrase (I, we, he or she or) is (you are) blessed (by God) implies that there are others who are deliberately not blessed by God. The common rejoinder is that every human being is blessed by God in their own unique way and simply must accept and understand the blessing God has bestowed upon each of them. Still this response fails to explain God's whimsy. The first problem is that God, as the Creator, has an equally vested interest in each human being.

Returning to our job hunt we had the one hundred applicants, who were all asking God to give them the job, which is, in reality, asking God to prevent the ninety-nine other people seeking the same job from finding work. God knows that the other ninety-nine may need the job just as badly or more so than the winning applicant but positively responds only to the entreaty of this one individual. Is the one individual special in the eyes of God? Were his or her prayers better? Is that individual just a better human being according to God's standards? And if God didn't specifically respond to that individual's entreaties He still, as omnipotent creator, altered the outcome in favor of any of the other job applicants.

Is all good fortune the result of God's blessing? What of the successful millionaire drug lord who orders the murder of dozens of people every year and causes the death of hundreds and thousands of others? Has God deliberately blessed him with his good fortune? And if God hasn't blessed him then how are we to interpret his status in this life which was created by God?

If God, being all-knowing, all powerful and all good, knows which of the failed applicants, in our aforementioned job search, will start taking drugs, or beat his wife, kill her children or commit suicide, then what is their blessing. Or did God deliberately withhold his blessing knowing the consequences. And if God is aware of these acts and is intimately involved in those details then it follows that God specifically decides who breaks a leg or cuts a finger. If that is true then either God has a reason for every minor injury and illness as well as every major catastrophe based on some unfathomable measure as to whom is in God's graces and who isn't or God is just playing games and allowing humans to suffer for His own amusement. Neither choice can be comforting to the religious who have faith that "God is good". Neither choice is reasonable no matter what you believe. What is the purpose of prayer or worship as a form of thanks to the Supreme Being? Does the drug lord communicate with God and thank Him for his blessing?

If, God is a conscious being of will and a source of only love, then evil has to come from another source and God could not exist as a supreme cause of all things because some other cause created evil which means there is another being the equal to God as a creator. Since that would contradict the very definition of God as the omnipotent Creator then any evil that occurs must be a creation of and caused by God. A supreme being possessed with a consciousness that creates and thus causes both evil and good must by definition be amoral. If God did not create or allow evil to exist then He cannot exist as the omniscient/omnipotent sole Creator and therefore could not by definition be God. While it is arguable that God is not the Creator, the Creator must be God. If there are two creators, one for good and one for evil, then neither is God.

God is an omnipotent creator, with a consciousness and by definition, controls every act and occurrence in the Cosmos. Every hurricane, disease, earthquake, and tsunami is the extension of God's creation, His consciousness and His decision to cause or, at the very least, allow, pain and suffering, and thus a product of His careful deliberation. Alternatively God has no motivation or interest in the consequences for His actions or failure to act and therefore is amoral, at least from a human point of view.

The inevitable counter argument is that God is not fathomable and therefore we do not have to, nor can we, understand anything that God does because that is the meaning of faith. Not surprisingly many of those, who contend God is unfathomable, also insist that they have communicated with God and know what He wants them to do, as in the phrase, "I asked God for guidance and He told me . . ." Unquestioning faith was the primary rationale of the 9-11 Islamists as well as most other Islamic jihadists who justify their murderous acts. Do these homicidal faithful simply misunderstand God's message? But by definition God knew the result of sending that message to those faithful so whether they misunderstood or not is irrelevant. Arguing that man has free will does not free an omniscient God from knowing the outcome of every act and therefore responsible for the results. And an omnipotent God has the power to change the outcome of any event. Of course if He doesn't have such power then the evidence of God's existence through miracles no longer exists for the faithful.

The argument that 9-11 occurred because man has free will while God, actively intervened and miraculously saving individuals from death in an incident He allowed to happen is incongruous.

God, the Creator, knowing exactly what the jihadists were going to do, and why, and also knowing the ultimate results of those acts, He intervened and saved the life of an individual (or two or three), whose life he put in jeopardy by not intervening sooner, and otherwise would have died, while simultaneously deciding not to save the life of three thousand other individuals.

So what is God's nature? Is God a rational being? Acting logically as demonstrated by the laws of physics and chemistry? Or is God irrational? Inflicting death and destruction on the innocent? Could an irrational God be all good? But how does a rational God destroy innocent lives while simultaneously physically awarding the lives of those human beings committing the most heinous of crimes?

High profile Christian ministers, such as Oral Roberts, Franklin Graham and Pat Robertson, (as well as several Imams and rabbis) have expressly stated that the traumas visited upon Haiti, Japan and even those killed in the 9/11 twin towers attack were revenge by the Lord for a myriad of transgressions (homosexuality being one of the most prevalently mentioned acts of evil) committed by a more general population. Yet none of them have implied or even hinted that the monetary awards reaped by Idi Amin, Stalin, or Hitler, which were a direct result of their murderous actions, were also provided by God. Were those national leaders and the thousands of Mafia leaders and other successful criminals able to enrich themselves and lead lives of luxury without God's approval or knowledge? Maybe the all knowing God failed to notice them or their acts of evil.

According to the rationale of a large portion of theistic leaders those acts, which these religious leaders characterize as most appalling to God, and thus worthy of his vengeance include, "unnatural" sexual acts (apparently not including pedophilia) but do not include, genocide, mass murder, or drug cartel activities, since the perpetrators of the latter crimes have often lived in luxury and security for much of their adult lives. Conversely, the millions of victims, none of whom affected anything remotely close to the Holocaust, the Cambodian death camps or the deliberate mass starvations perpetrated by Stalin, were logically and rationally killed and tortured with God's full knowledge by the very men who reaped His earthly benefits.

Thomas Aquinas stated that "Whatever good we attribute to creatures, pre-exists in God". Aquinas does not address the question that naturally follows: 'Does whatever evil we attribute to creatures, pre-exist in God? To answer that in the negative results in God not having any control of evil and therefore evil exists independent of God which is anathema to an omnipotent being/creator. By definition nothing can exist outside the control of a creator. If God is the Creator then either God created evil or evil was created independent of God and therefore negating God's position as an omnipotent creator and inferring that a second equally omnipotent creator exists, an evil equivalent of God, which would contradict the definition of God. The only conclusion is that if God is the Creator, He knowingly and deliberately created all that is good and all that is evil with the full knowledge of the consequences of His creation.

Alternatively, evil must be a separate creation of the devil, who was originally an archangel, whom God created. Either the Creator created everything and knew all the consequences of that creation or as omnipotent had the power to change the outcome. But the omnipotent Creator not only allowed the Devil to create evil but to foment its growth. The troubling question remains that if Lucifer and Satan, are all evil, why and how did they emanate from the infinitely good Supreme Being?

Or is evil solely the fault of humans' failing to submit to God's will? That, immediately after creating human beings, God chose to tempt them in order that they prove to Him that they could be trusted and faithful? For what purpose does the omniscient/omnipotent/infinitely good Creator need to test the character of an inferior being, already knowing the results of that test? Such a test of will would seem more suited for the insecurities and suspicions of the Godfather rather than God the Father.

Is evil just a matter of choice? But the real question is what is evil? The Shoah, the Killing Fields, Stalin's purges, the Armenian and the Rwandan genocides are all universally classified without hesitation as evil because they resulted in the death and suffering of millions of innocent human beings, within just the last century. What of the deaths and suffering of millions of innocent human beings as the result of earthquakes, hurricanes, and other natural disasters, including epidemic diseases? Why are those results not considered evil if both categories of mass death and suffering are incurred with the conscious forethought and knowledge of the Creator? Is it less evil when a mother watches a child die slowly and painfully from AIDS or Ebola than if the child is shot and dies instantly because the act of shooting the child is done by a human being while the slow painful death from disease is considered an act of God? If the answer is yes then the faithful are stating that God is not omnipotent and can't intervene. But those same faithful will unhesitantly thank the Creator for his intervention if their child survives either the shooting or the disease. The only conclusion is that God decided to intercede in one instance and not in the other. God had the power to intervene in one and deliberately failed to do so by allowing the evil, either directly through human acts or indirectly through disease which He created.

According to at least some fundamentalists, God punishes humans, innocent and guilty alike, with death by allowing natural or man-made disasters while, simultaneously, according to these same fundamentalists, awarding the innocents with entrance to eternal life in heaven. Even the most ardent fundamentalists will testify that many of the humans that died on September 11, 2001 or in the tsunamis which have killed tens of thousands in Japan and Southern Asia, in the last decade, were innocent children and others who may have been good Christians or Muslims or Buddhists and worthy of heaven. So it would be just as reasonable to assume that the so called punishments for these Earthly sins are actually rewards. But that would contradict the fundamentalists who preach that disasters are God-made punishments. It follows that God has allowed thousands of deserving souls to enter heaven earlier than the rest of us by killing them in an earthquake, tsunami or terrorist attack. In which case human condemnation of murder by capital punishment is terribly misplaced and murderers should be rewarded for their generosity for allowing more humans, their victims, to be early bird entries into the Kingdom of God. Logically, murder is just God's way of using human beings, rather than natural events, to bless innocent victims with Heaven. Unfortunately, the blessing of heaven for innocent victims fails to explain the suffering of tens of millions who remain alive and must live with horrific mental or physical trauma not to forget the tens of millions of their family and friends who suffer with them.

If on 9/11 God was punishing human beings for homosexuality and other moral transgressions isn't he perfect and powerful enough to have punished only the moral deviants in any of these disasters? Did God really get confused and mistakenly include some ardent homophobics who preach against homosexual behavior when he visited death on the twin towers? Counter argument—the homophobics were martyrs for God's will.

The typical priest, minister, imam or rabbi loves to tell the mourners at a funeral that the victim of the accident, illness, murder, etc., is actually being rewarded for his good life. It logically follows then that a murderer is allowing somebody to enter heaven sooner and is therefore doing a good thing. But many of these same religious support the death penalty as the penultimate punishment. The ultimate punishment we can assume is an eternity in hell assuming all of these people who receive the death penalty must be going to hell. Is it such a sure thing that every guilty person that receives the death penalty is going to hell? Isn't it possible that God would forgive a few of them? In which case the point of the death penalty is moot and in fact acts as a reward.

Once a person dies and their soul learns that God actually exists and then becomes sincerely and deeply repentant shouldn't the infinitely good God forgive their sins, which Jesus died for, and allow them entrance to heaven or at least eventually after some time in purgatory?

Are these actions of reward and punishment meted out so randomly the result of careful considerations of a logical and rational God? Or are we to conclude that God is neither rational nor logical? Or does God only get involved when he feels like it? If God awards randomly then what is the purpose of prayer? Einstein's argument that God does not role dice with the universe is clearly suspect considering Einstein's circumstances as a Jewish ex-patriot hounded out of Germany by arguably the most evil government in the history of the world. Does God have a dual nature which simultaneously acts capriciously and rationally? The George Burns' version of God in the movie "Oh God" telling John Denver that He doesn't get personally involved because He has provided humans with everything they need and it's up to them to do the best they can would seem far more plausible but doesn't resolve the natural disaster conundrum.

But can an anthropomorphic omniscient omnipotent creator be a non-involved entity? If God consciously created the cosmos then from that point on God is responsible, by definition, for everything that happens. If God consciously responds to any human entreaty, resulting in a miracle, then the failure to respond to any other entreaty is also an equally conscious act. Therefore, any occurrence, which affects any form of life, is the result of a God's conscious, knowing, and deliberate decision. Whether it is the act of falling asleep at the wheel and accidentally killing an oncoming driver, deliberately killing thousands of people with one bomb or an act of nature like a tsunami, God is responsible. The death camps in Germany and Cambodia, the fourteenth century black plague and the influenza epidemic of 1918, starvation of children in central and southern Africa and Bangladesh are the result of cognizant decisions, not to intervene, made by the Creator.

If God is personally and intimately involved with the outcome of golf, football and baseball games (He must be according to the athletes who thank Him) He must equally be actively involved with the death of innocents. Or, in a deistic alternative, He is at least deliberately ignoring the pleas and prayers of the millions of people each year who die in floods, famines, earthquakes, and epidemics which are deliberate acts by God. If He hears the prayers of the fans cheering for their favorite team winning the Super Bowl He must hear the prayers of the children and their parents swept away in a tsunami. If God is responsible for the miracle cure He must be responsible for the disease. If God is answering all these prayers with rewards and punishment as He sees fit at that instant in time, then it would be logical to conclude that God is not just infinitely good but infinitely good and evil. Why must God be perfect only in terms of good? If God is perfect shouldn't perfection include evil?

Evil can be measured through a direct one to one correlation with the intentional pain and suffering that is caused by the person inflicting the harm on an innocent victim. Can evil be measured when there is unintended pain and suffering? According to Part III Section One, Chapter I, Article 4, paragraph 1759 of the Roman Catholic Catechism "An evil action cannot be justified by reference to a good intention" (cf. St. Thomas Aquinas, *Dec. praec.* 6). Presumably the Creator had good intentions in providing the natural laws of science that result in hurricanes, tsunamis, earthquakes, the Black Plague, AIDS and a plethora of other acts that caused and have caused pain and suffering. Biblically God is credited for the Flood, the Egyptian plagues including the death of first born males, the drowning of the Egyptian army, David killing Goliath and Samson's murder of hundreds among dozens of other intentionally inflicted acts of harm.

According to many Biblical or Koranic theologians God's actions are justified because He is God and we should do as He says not as He does. God has the right to play God and can take human life or cause pain and suffering on a whim because He is the one Who created the life. This whimsical God, by this reasoning, is an amoral being, Who demands moral behavior. Faith in this Creator provides total justification for a human being to pursue any act of evil by stating that they were acting pursuant to God's wishes because God's wishes are always for the good.

While it is easy to single out suicidal jihadists, the same justification is used by every people, tribe, nation or religion in the history of the world engaged in any type of warfare. Every participant in every war has fought in the name of God except for the few very confident atheists. Every German, Japanese, American and Russian in the Second World War had God on his side, soldiers and civilians alike. Every Catholic, Protestant, Jew and Moslem had

God on his side in sectarian wars past and present. No soldier has ever participated in a battle and shot and killed or bombed their enemy without faith that they were acting not only consistent with God's will but as His instrument. If there is a Creator, then every one of those soldiers must be correct.

If God, the Creator, can take life or inflict pain and suffering, using whatever instrument is available (natural or human) then no act that inflicts death, pain or suffering can be classified as evil since each instrument is, in effect, an extension of God. A creator God eliminates any difference between moral and immoral behavior since any act can be explained as pursuant to God's plan as the Creator. Whether the instrument is an unbalanced teenager shooting 26 school children, a life-long psychological disability, or a Mafia hit-man just doing his job, each is an instrument of the Creator and a part of His plan.

The Catholic Encyclopedia defines Evil in part as: "including all that causes harm to man, whether by bodily injury, by thwarting his natural desires, or by preventing the full development of his powers, either in the order of nature directly, or through the various social conditions under which mankind naturally exists. Physical evils directly due to nature are sickness, accident, death, etc. Poverty, oppression, and some forms of disease are instances of evil arising from imperfect social organization. Mental suffering, such as anxiety, disappointment, and remorse, and the limitation of intelligence which prevents human beings from attaining to the full comprehension of their environment, are congenital forms of evil; each varies in character and degree according to natural disposition and social circumstances."

If God is the Creator then everything that occurs in the Cosmos is part of God's plan, then either there is no act that is evil or a part of God's plan is evil. How can any act be found evil if it is part of the Creator's plan and the Creator is infinitely good? There is either a co-creator or the infinitely good Creator also created evil. By definition God cannot be a co-creator. Therefore God the Creator must have created everything that is evil since anything that exists cannot exist unless God created it, if God is the Creator.

Most, theists who declare God as the Creator excuse God from the creation of evil. Instead they reason that God allowed evil in order to test man's free will. While this explains man's evil acts it does not answer the evil results of natural scientific causes which fall under the definition of evil quoted above, but have no connection to freewill. And if God did allow evil in order to test man's freewill then evil is clearly part of God's plan at the time of Creation. In fact according to Catholic Doctrine, and most other theological defenses of the existence of evil, evil is absolutely necessary in God's plan so that humans can demonstrate their devotion to God by choosing God over evil.

The problem with putting a buffer between God and the existence of evil by proclaiming that he merely "allowed" evil to exist rather than created evil evades the dilemma that God knowingly created the angel whom He knew would become Satan or Lucifer. Or, in lieu of belief in a superior evil being, he created the humans who have chosen evil with His foreknowledge as the omniscient creator. In either case all evil that exists was created by God. Evil cannot exist outside of God's plan if God is the Creator. Good cannot exist outside of God's plan as the Creator. Every act of good can only be measured against a comparable act of evil. An infinitely good God can only be measured against an infinitely evil God. God the Creator can only exist with a perfect all encompassing nature.

Only if God is not the Creator can acts of evil be the sole responsibility of human beings. Only if God is not the Creator can natural scientific forces that result in death, pain and suffering occur absent the specter of evil. Absent a Creator, no person of faith can excuse or rationalize their behavior as part of God's plan. Absent a Creator, humans cannot blame God for things that happen as the result of biology, chemistry or physics. Only a non-creator God allows for true free will and forces all of mankind to take the ultimate responsibility for their actions because if He is not the Creator, God cannot have a plan.

But if God is not the Creator, can God still intervene pursuant to our prayers as "our father"?

3

The Case Against God
the Father

IN THE OPENING SCENES of "The Godfather" Don Corleone is grant-
ing favors to people who have either previously demonstrated their
devotion or promise to demonstrate their devotion. All of these
"friends" of the Godfather have tried every other available avenue
and failed to get the help they wish. The Godfather is their last
chance and each is awarded for his or her personally sworn devotion through the Godfather's personal intervention. Later we see the
Godfather taking brutal retribution against those who have rejected
his friendship by acting against him or failing to support him.

It is not just the fundamentalists but the general population
who adhere to the belief that God is involved in the minutiae of
our lives. Except for the most ardent atheists, everyone in a time
of personal crisis will pray to God for divine intervention. The
expression "There are no atheists in foxholes" is probably more
accurate than agnostics and atheists would admit. On an everyday
basis the expressions such as "Oh my God (OMG)", "Thank God",
"Goddamn", and "What in God's name?" are thoughtlessly used by
all in every culture.

The belief in a personally involved God is not only comforting but the primary rationale for proselytizing and sustaining the

faith. Whether it is Jews, or Mormons, or Catholics, or Protestants, or Muslims, or Animists the message is communicated that God will protect you from harm, heal your wounds, save you from disaster, ad infinitum, if you just ask. The most religious, the most faithful, attribute God's personal intervention for their good fortune without acknowledging in any manner that God's original plan put them in them in the unfortunate position from which, He then extricated them at their request. Conversely, asking God to change His plan, demonstrates a total lack of trust in God. Praying for divine intervention, is saying to God, "I don't like your original plan. Would you please change it for my personal benefit?" Furthermore, those who are most likely to beseech God the Father to intercede on their behalf, are the same people who extol the rest of the world to have total faith in God and His plan while the atheists and agnostics are more likely to just accept the consequences of God's plan without asking for personal self-serving intercession. These same faithful and devoted also believe in God's retribution against those who have rejected his love failing to demonstrate their devotion.

The faithful turn to God the Father to confess and ask for forgiveness for actions which are part of the omniscient being's plan. But, if God has a plan then everything that happens is part of that plan and there is nothing that requires forgiveness. If God does not have a plan then anything that happens is out of His control that would contradict the definition of an omnipotent and omniscient God the Father Almighty Creator of Heaven and Earth.

If God the Father has no plan, we are left with a Father Who capriciously allows or instigates pain and suffering among millions of the sincerely faithful every year. If God the Father has a plan, then we have a Father Who intentionally inflicts the same while responding, or not, to prayers for his intercession.

Would Hitler, Pol Pot, leaders of drug cartels, or Mafia hit men be wrong, in thanking God the Father for their good fortune? God the Father allows or bestows, and has allowed or bestowed, upon the most evil human beings, the most fabulous wealth and its accoutrements imaginable. Everyday billions of people formally thank God for food on their plate and a dozen immaterial "blessings". If a palatial home on a huge estate with expensive automobiles, private jets and designer clothes, all purchased with profits from the sale of cocaine and heroin and caused the death of hundreds, including murder, are not a blessing from God the Father Almighty, then who is responsible for bestowing these blessings? If the answer is the devil then the admission is that the devil is at the very least, equally almighty and powerful as God, which would negate God the Father as omnipotent. By definition a being who is not omnipotent cannot be God. Therefore there is no possibility that the devil can be as powerful as God and able to bestow blessings from an evil source without, at minimum, God's approval. The only other rationale is that the riches and good fortune of drug-lords, malevolent dictators, and other similarly situated individuals is that their "blessings" are part of God's plan and thus the only answer to the question posed is that Hitler et al., are right in thanking God the Father for their good fortune.

Clearly everyone has faith that God the Father plays favorites. This, by the fundamentalists' own teachings, is the same God who loves everybody equally. Jews, Muslims, Christians, and their plethora of sects, fervently believe that they are number one in the eyes of God. And many of the most sincere believe that God has individuals whom he favors over others. Christian, American athletes are always proclaiming their gratefulness to God and Jesus for helping them win the game and of course didn't do anything to help the opposition. Couldn't that team had won if God hadn't helped. Why did God help that team more than the other team? Does God only help winners? For the last nearly ninety years it is clear that God is a major New York Yankee fan but absolutely despised the Chicago Cubs. When one of these athletes stands up after a game that his team lost and blames God for refusing to help them then we will have seen a true religious experience. "I want to thank my Lord God and Jesus for helping the winning team beat us today, even though we prayed harder than they did."

The same may be held true of those who thank God for their ongoing health or for their recovery from a nearly fatal illness.

Shouldn't they be thanking God for the excruciating pain and suffering caused by a malignant tumor? The only documentation of such an act is that of St. Bernadette who (at least on film), despite fatal and incredibly painful tumors covering her body continued to thank God for His blessings.

If God's blessings are totally indiscriminate does that mean that God is not perfect? Does God have to be perfect in the sense of being infinitely good? The Greek and Roman gods were very imperfect with the most human foibles. Egyptian and Norse gods were similarly defective. The Jewish/ Muslim/ Christian God is perfect by definition. Can torture and acts of depravity be the result of a perfect plan? Can a perfect all good creator create anything less than perfect? The problem may not be the definition of God as perfect but the definition of perfect as "entirely without any flaws or shortcomings, faultless." (Oxford-English Dictionary) God that omniscient/ omnipotent Father Almighty is conceivable because that perfect God must include all that is evil in order to be absolute and complete and consistent with the Cosmos as it exists and has existed. If anything exists but cannot be ascribed to God that makes God less than absolute and complete. If God the Father is absolute and complete then He must not only be infinitely good but must be infinitely evil.

The Biblical argument is of course that God didn't "create" evil because that was the free choice of—Lucifer, Satan, Cain, etc.—Then God must have allowed evil to happen. The Biblical anthropomorphic God certainly has never prevented evil either by the hands and minds of humans or natural disasters including pandemic diseases. Does the Biblical God have no control of evil? This would contradict the very premise of God as the creator. An omniscient/omnipotent God must, by definition, encompass all the evil in the Cosmos because the Cosmos constitutes a perfect all-inclusive God. God is malignant and benign. Alternatively, the argument is that Lucifer, Satan et al., have power equivalent to the Biblical/Koranic creator because they (or whomever) have the ability, superior to God, to induce evil thoughts and/or actions. Therefore the only conclusion is that God created a more powerful being than Himself. So either God allows evil or God created evil or God is too weak to combat evil. A supreme being that is not omnipotent and omniscient cannot be defined as God.

We objectively describe the behavior of all other animals without judging motives or their proclivity for evil or goodness. We watch animals kill other animals in the most brutal manner without any thought of whether they will be awarded or punished in an afterlife. Killer whales toss helpless sea lions around like beach balls. Lions calmly eat a still breathing zebra. Male lions kill the cubs of rival males. Elephants trample people in a hormonal rage. Even domesticated dogs are not categorized as evil and condemned to hell when they kill a small child. Chimpanzees kill other chimpanzees. All these animals are forgiven their behavior because they are acting like what they are. It is their nature. None of their actions are considered evil as observed objectively and non-judgmentally by human beings.

An objective and nonjudgmental observation of human behavior by a non-Earth being, say from the Andromeda galaxy, would certainly include the following observations:

> Humans have always physically attacked and often killed other humans. Always! It is clearly their nature. These attacks can be as an individual against another individual sometimes done randomly and sometimes as part of a strategy to attain power over the other individual. Very often these attacks are executed as part of an organized group or tribe with whom the human associates but always to attain power over a rival group or tribe in what appears to be done in a formal ritual.
>
> Curiously, most of those humans, apparently believe, without any evidence, in a supreme being, a Creator, they call God, and Who, they believe, metes out an eternal punishment as a direct result of their behavior after their own death. Humans also believe that attacking and killing other humans will result in the eternal punishment by the being they call God, but are unable to control their own individual actions which will, according to their own understanding, cause them to suffer eternal punishment. Humans have no natural enemies, which

35

they need to fear other than other humans. It is also odd that humans clearly understand that attacking other humans is bad for their species, except in self-defense, and will verbally and physically condemn another human for attacks that cause injury and death, but have continued to attack and kill other humans for tens of thousands of years despite their understanding of the consequences to their own species and consistent with their beliefs in the being they call God.

And, if by chance, these Andromedans, also ascribed to a Creator God the Father, then they might also include in their report that Humans apparently understand the difference between good and evil, but persist on acting out against other humans with full comprehension of the consequences from both other humans and from God. Any conclusion from our Andromedan visitor is more likely to utilize adjectives such as "intellectually challenged" rather than "evil". If they observe us closely they might also notice, ironically, that those humans that are the most "intellectually challenged" are the least likely to deliberately and with complete understanding of the consequences of their actions, attack or harm other human beings.

Only a subjective observation of human behavior, by other humans, uses the term "evil" or "good" to describe any type of human behavior. Humans, when observing and describing behavior of other species, never use the terms good or evil. Behavior that, if it occurred between two humans, cleaning each other, protecting each other, feeding each other would be judged as good things are expressed as merely acts which are in the best interests of promulgating the species. Species that act out in ways that humans would consider evil are just doing what is best for their species. Is it possible, objectively, that acts that are subjectively evil or immoral or depraved might in fact be what in the long run is best for our species. Despite, or as a result of, all the wars, murders and other apparently detrimental interactions between humans, the human species is more populous and generally better off physically (fewer diseases, longer life-spans, taller and heavier etc.) than any other time in history. If we judged our species by the same criteria we judge zebras, lions or whales, we would have to conclude humans are thriving as a result of their interactions which have strengthened the species.

FIGHTING FOR GOD THE FATHER

No war has ever been fought to preserve and promote atheism. Modern day atheism is an ironic by-product of Communistic political values since the most communistic peoples of the world are nuns and other monastics. No Communist country ever entered a war because it wanted to impose atheism as its primary motivation. Russia and China instituted communism as a protest against economic systems which favored an oligarchy. Ho Chi Minh wanted to unite Vietnam and Communism was the vehicle not the motivation. Ditto the Korean War. Wars between religious philosophies fought in the name of a deistic religion are eternal. Muslims and Christians have historically fought other cultures for the primary purpose of imposing their religious beliefs on another people. Crusaders and jihadists are acting with the exact same belief that the blessing of God is an absolute, as a reward and motivation for their acts and which, of course, contradict the teachings of their own theology.

The perpetrators of jihads and crusades conveniently interpret their belief in the Ten Commandments a little more loosely for themselves. Thou shalt not kill except when forcing others to believe in God in the same manner as you do. Compelling people by physical or psychological means in order to demonstrate their beliefs in God in a manner acceptable to other Islamic jihadists or Christian crusaders in order to please God is convoluted at every level.

Why does a Jihadist fight for Allah? Why does anybody have to fight for God? The U.S. military makes it a point to fight for God and country. If one believes in an all powerful creator why would He need anybody need to fight for him? Are they afraid God might lose? And if God loses who would win? Can God lose? If a Jihadist or the U.S. loses does God lose? What does He lose?

Does God need defenders because He can't defend Himself? This is the definition of hubris: a single member of a species of seven billion, on a small planet, in a minor solar system, in an average sized galaxy, which is one of billions of galaxies in the universe, thinks he is fighting to defend the Father Almighty and Creator of that universe.

If God actually exists does the Father Almighty and Creator of the Universe really need defending? Defending Him from what? Is there anything that can harm God? God's actual existence needs no defense. Then what is meant by fighting to defend God is that one is fighting defend one's own personal concept of God. Everyone fighting a war to defend one's concept of God are all fighting to defend the same God, because all the adversaries believe there is only one God. The adversaries are therefore only fighting because they believe that any practice of worshipping the Father almighty, other than the one they practice, is offensive to God.

4

The Case Against an Afterlife

MY MOTHER LOVED TO tell the joke about the bus that crashed killing everybody aboard so that they all arrived in Heaven at the same time and are taken on a tour by St. Peter. First he guides them to a beautiful church and tells them this is the Mormon Temple and the tourists "oooh" and "ahhh" over its overwhelming and spectacular beauty. Next they approach a Jewish Synagogue and again the tourists are awestruck by its architecture and elegance "They really are the chosen ones". Next they stop at a Mosque and are taken aback by its breathtaking magnificence, "Allah truly is great". Then, as they approach an exquisite Cathedral with long colorful windows, St. Peter puts his hands to his lips and hushes his tourist group. Next they stop in front of an enormous baroque style church with spectacular flying buttresses. As St. Peter begins to explain that this is the Anglican church he is interrupted by one of his guests.

"Excuse me sir, but you didn't tell us who was in the Cathedral we just went by and why we had to be silent."

"Oh, those are the Catholics and they think they're the only ones up here."

I think most of the Christian sects and followers of Islam could tell the same joke. If there is a "heaven" of some kind it would seem that the gatekeeper should be more inclined to grant entrance to the atheists than to the Popes, Ministers, Ayatollahs, or Rabbis, because atheists always act without an ulterior spiritual motive of seeking a reward from God for performing any good works. Those who believe in God act either out of fear of punishment or in pursuit of an eternal reward. An atheist's good deed is done purely because he believes, at the time; it was the right thing to do. This underlines the fact that if atheistic views prove wrong that God would be more likely to accept totally altruistic humanitarian acts than those of the faithful, who are looking, or at least hoping, for a reward for performing the same good deeds. The faithful are merely seeking an answer to the question "What can I do to please You in order to be rewarded heaven or to avoid hell?"

Theists, who believe in an afterlife, teach the necessity of good works either from a perception of bribery or blackmail. Either one acts to be rewarded with eternal happiness in heaven, or one acts to avoid the eternal damnation of hell. The more religiously fundamental the individual the greater the perception of the bribe or the blackmail. A true atheist acts for totally altruistic reasons. He or she is neither in fear of the fires of Hell nor longing for the serenity of Heaven. The motives of the Atheist or a Hindi performing good works are, by definition, far purer than the theistic individual, because the atheist and Hindi, acts because they believe it is the right thing to do as a human being without an after-death consequence. Ironically, if you believe that Jesus was the human incarnation of God, then you must believe that Jesus was the only human being who acted without any motive of being rewarded with heaven or punished with hell.

How does the almighty Father decide who will go where? Are there just so many sins before one can no longer return into the good graces of God? Is there a specific sin from which there is no chance of forgiveness? Are there any cases of any human being who we know has consciously acted to displease God? There is no record that supports the hypothesis that Hitler didn't believe in God and wasn't sincerely acting on His behalf. Does Hitler or the

9/11 jihadist get judged and condemned to hell by God because they got it all wrong despite sincere beliefs to the contrary?

When I was sixteen I got into a heated discussion with the John Whelan, the Archbishop of Hartford, concerning my Jewish relatives and their eventual, if ever, entrance into Heaven. After conceding that they were very worthy, by their acts and suffering, and maybe worthier than a lot of Catholics, Bishop Whelan ended the discussion with the rhetorical question, "Would you rather take a rowboat or a jetliner to Europe?" regardless of their goodness and good deeds, their souls' unlikely entrance through the pearly gates would be greatly delayed or at worst, in danger of being swamped and lost in a purgatorial sea. In Whelan's view, faith trumped actions. This seems to be the rationale for Koranic/Biblical fundamentalists in order to excuse any sin that they themselves might commit. The logical but cynical, and admittedly sophomoric conclusion, according to Archbishop John Whelan, is that because of Hitler's Roman Catholic upbringing, God the Father would allow him to advance to heaven, ahead of the six million Jews whom Hitler exterminated.

The threat of hell and promise of heaven are prime motivating factors in the behavior of the faithful. And at least for those who believe that Jesus was God incarnate, what was Jesus motive for His good deeds, since He knew that He could be neither rewarded nor punished with heaven or hell? An atheist, who acts with goodwill to his or her fellow human beings, would seem to have much more in common with Jesus than anybody whose constant thought is how to most effectively avoid hell or enter heaven. Every theistic faith has a slightly different set of rules, which they claim God the Father uses to execute eternal damnation or eternal happiness. The commonality of the rules is that God the Father judges each human based upon the morals of the theology that is making up the rules.

The very concept of hell itself is diametrically opposed to a God who sent His son, Himself incarnate, to our world to sacrifice for our sins. Did He die just for a particular group of sinners or a specific number of sins? There is no time in history where man has not acted in the most atrocious and despicable manner. The empirical evidence strongly suggests that man's acts of evil are nothing more than the manifestation of the nature of human beings. No different than a shark or a lion or a killer bee. Thus the concept of eternal damnation is meaningless in that God would not be expected to punish any creature for acting in accordance to his nature. For if man is naturally flawed then for what purpose would God inflict punishment?

On the simplest and most practical level how does God the Father decide who is condemned to hell? There is no reason to believe that sinners are not statistically on a bell curve and there is a continuum of sinfulness. Does God decide that one person with "N" number of sins goes to hell and the person with "N-1" sins goes to heaven? Or is His decision based on the quality of sin which is also on a continuum. If one sin is just a fraction more immoral than the other sin does that condemn the unlucky sinner to eternal damnation although that sinner didn't know the cut-off point in advance? And if all of God's creations and their actions are part of His plan how can any of them be suffer condemnation for acting within God's plan? Can anybody act outside of God's plan? The alternative is that there is no plan and all things are random and outside of God's scope and power which would negate His omniscient/omnipotent theistic status.

Why is Heaven a reward or hell a condemnation? Aren't the rewards and tribulations met during this lifetime sufficient? Why should God allow a form of life to continue eternally? Isn't the life provided reward within its own limits? If God the Father Almighty and Creator gave us this existence shouldn't we be grateful for that gift? A reward of Heaven, as conceived by humans means that they believe the gift of life was insufficient.

If one possesses a soul which is awarded or condemned in an afterlife that infers that the soul has an awareness of its surroundings. An awareness of anything requires the engagement of the brain. The engagement of the brain requires the flow of blood carrying oxygen. Therefore, in order for the soul to derive the benefit of an afterlife it requires a re-formation of the brain and circulatory/

pulmonary system so that the soul can appreciate its heavenly (or hellish) surroundings. So the same Creator Who provided the laws of nature which formed the entire cosmos must resort to magic for the creation of this parallel universe of heaven and hell.

Most human beings seem to be drawn to a God they can worship and believe in as a creator. Worshipping an anthropomorphic God by means of prayer, ritual, and sacrifice are, in their essence, the means to attain a place in some form of the perfect afterlife, heaven, by currying favor with the anthropomorphic God. Such acts are done more as an effort to avoid the eternal torture to which God condemns those, who do not reach some nebulous level of prayer, ritual and sacrifice. The criteria for the appropriate level and forms of worship are set by the very specific form of worship to which a human being may be religiously attached. In fact the exact same ritual which is the ultimate form of worship in one religion may be the greatest blasphemy and sure fire condemnation in another. Fundamentalist Christians adorn their homes with pictures of God and Jesus. Fundamentalist Moslems issue fatwas and eternal condemnation for displaying pictures of God and Muhammad. Both fervently believe they are righteous to do so in God's eyes and have fought and continue to fight to prove they are God's Chosen people.

THE CASE AGAINST HEAVEN AND HELL IN THE COSMOS

Taking a round number, there have been to this date very approximately 100 billion people who have existed, or will exist, on the planet; let us then assume that those 100 billion people's morality can be scaled from 1-100 on the "Evil Meter". Let us also assume that the results of our "Evil Meter" would fit nicely on a Bell Curve. And even if you don't like the Bell Curve it is I think arguable that Evil can be measured on a continuum among our 100 billion human beings. If there is a hell in the cosmos where some percent of those 100 billion will reside for eternity then those who prescribe to a hell also prescribe to the theory that some separate consciousness is going to place a clear divide between two per cent of those 100 billion and make the judgment that human number 1 billion (where the last one percent of the population falls on our evil meter) deserve to be in hell for eternity while 99,000,000,000 are spending eternity in heaven. Can a clear line be fairly drawn by a God who isn't capricious? Can a just and fair Supreme Being distinguish between this much evil and that much evil at any point on the spectrum? Is evilness quantified or qualified? Is it how many times one sins or the kind of sin that is committed? Can a true believer be continually sinful because he believes what he does is for God? Is a true atheist forbidden from Heaven regardless of his always positive acts?

Those who believe in Satan (or Lucifer, Beelzebub, etc.) also believe in God as the creator of everything yet refuse to accept the fact that God must then have created all that is evil including Satan. If Satan, et al., are in control of evil does that mean that they are more powerful than the being that created them? Is it reasonable that God created an equal rival? If God didn't create Satan and all that is evil, who did? A second creator?

Let's remove the Satan persona from the equation since it is the weakest argument, on its face. It is often argued that while God created evil God doesn't control evil or the acts of those who do evil things, which brings us back to Nature. According to theistic theory Nature is totally under the control of God. Even under the man made laws, whether theist or atheist, the term "an act of nature" is universally interchangeable with "an act of God" in many insurance contracts.

Do both man's freedom of choice and natural law fall outside God's control? For if God is in control of nature then He is by definition responsible for the indiscriminate killing of tens of thousands of innocent victims every year which, cumulatively, is more than all the deliberate evil acts of all the human beings committed, through freedom of choice, in history. If either freedom of choice or the laws of nature fall outside of God's control then what of those acts for which we thank him for his blessing for saving us from death, injury, hunger, sorrow etc? Often these blessings are the result of the same natural disaster or deliberate human act. How many thanked God intervening to save their lives in the 9-11 attack and the Indian Ocean Tsunami? So God is consciously either allowing or perpetrating an immoral occurrence in both instances (indiscriminate killing) while simultaneously acting, just as indiscriminately, to save thousands of others. Is the prayer to our creator then "Thank you God for not torturing, killing or maiming me (or my child, spouse, etc.)"?

The faithful must believe that the same consciousness which created the laws of nature which so reasonably explains the development of our species, as well as why a ball falls to Earth when it is dropped, and exactly why and how an atomic reaction occurs is the same conscious which irrationally and capriciously and deliberately causes death and destruction. That irrational Koranic/Biblical supreme being is then essentially no different from the Roman/Greek god Zeus/Jupiter, or countless other human like gods that exist so many cultures, and suffer and express all the emotions of any human.

It is complete utter and total nonsense to attribute to a being, who one believes to be the all powerful, all good, all loving creator of the universe, an inability to control primitive human emotions or to act in anything but the most rational and reasonable manner. In religious terms it can only be sacrilegious and blasphemous to attribute to God (anger, vengeance, loving etc.) any kind of human emotions on which He would act childishly without considering the consequences of all those who have to be affected. Attributing to God the same qualities as man makes Him petty and His decisions inconsequential.

God isn't evil. God isn't good. God is. God is perfect if you use the definition of perfect as complete, absolute, total and encompassing everything. What is the reasonable alternative to a supreme being including all the evil in the cosmos? The Biblical explanation is that another being, the devil, (Lucifer or Satan) controls and created the evil. Obviously this explanation contradicts God as creator. God either created everything including Lucifer and Satan and thus all evil or God is only one of two creators which the Bible faithful reasonably rejects. Therefore, according to the most fundamental fundamentalist, God, by Biblical definition, created evil.

An appealing argument for bad things happening to humans is that God knows but does not interfere. God knew that humans would evolve to be the species we are: loving, caring, murderous, gluttonous, conniving, hypocritical, sincere, intelligent, idiotic, egocentric, altruistic, frightened, and courageous. It isn't a question of what God wanted or didn't want to happen but as in today's lexicon, it is what it is. This is God as creator but not administrator or bureaucrat. Yet there is still the argument that the omniscient creator of natural law has knowingly allowed the destruction of beings which were created through the natural laws which He created. And if God is omniscient then His knowledge of the pain and suffering which He allows puts Him in the same position as that of a Supreme Don Corleone. God as Creator, the Father Almighty, has blood on His hands. Like the Godfather, the Biblical/Koranic God expects unwavering loyalty. In return for allowing us to live normal lives he wants us to pay tribute. Sometimes he punishes those over whom He has dominion, sometimes He doesn't. The parallels are infinite. This is the nature and form of God which seems most reasonably rejected by atheists. For an atheist to reject that a Creator exists because the Creator is little more than an indiscriminate killer doesn't seem at all unreasonable.

Just as we anthropomorphize God so we anthropomorphize evil. The fact that things happen, which we characterize as evil from our viewpoints as human beings, does not mean that those events, or beings, are objectively evil. Evil means, pragmatically, that something bad or unfortunate has happened to human beings, that is morally reprehensible, caused, directly or indirectly, by another human or an act of nature, including wild animals and natural disasters. But is there evil that exists in the cosmos objectively? Is any act reprehensibly immoral when perceived from an objective view of a being outside of human perception? Would anybody find the acts of an army ant colony destroying or attacking other animals reprehensibly immoral? Are lions immoral when they kill their young? When a black widow or praying mantis eats its sexual partner? (Although this last example may have some highly moral support) If those natural acts were judged by human standards of behavior the label of evil is automatically applied. If there were a being perceiving our acts objectively would that being judge our acts as evil? The tsunami, lion, praying mantis and Jeffery Dahmer are all Biblically and Koranically creations of God causing death. Judged objectively, absent a human form of values, which is necessarily narrow in its concept, all of them are evil. But, if those acts are perceived objectively, they are merely the natural and innate acts of those particular Biblical/Koranic creations of God.

Any Koranic/Biblical or other theistic argument which insists that God, by definition omnipotent and exists as a separate conscious entity must accept that that separate consciousness is directly responsible for all immoral, evil and destructive acts by either instigating them or allowing them to occur with knowledge of forethought. The callousness of such a Being is inconsistent with any Koranic/Biblical or similar theology of a forgiving God or reason for humans to act morally. Arguably the mythology of the Greek, Roman and Norse Gods would be very consistent with this form of a consciousness possessed by a supreme Being. To allocate to God any sort of separate anthropomorphic consciousness creates a supreme being of small minded pettiness devoid, or at best, negating of any qualities that would be considered superior to human beings.

As to Heaven as an eternal reward for living a life that is acceptably moral according to some separate supreme consciousness the existence of the soul following the physical death of a human being, then the soul, as it exists without the physical restrictions of a human being and "heaven" is meaningless.

Heaven seems more likely to exist in the string theory system that includes parallel universes because heaven could exist in its own parallel universe. Raised and inculcated as a Catholic it is difficult to separate myself from the idea of the Resurrection, the soul and heaven and hell. Jesus is an historical fact. His nature is a matter of faith. Jesus' philosophy of life is inscrutable. But belief in his nature as the Son of God and his ultimate resurrection does not seem necessary to fulfill the objectives one requires to be a good human being. 1.5 billion people believe in Jesus, God reincarnate.

Does that mean the other 6 billion are going to hell? The belief in some kind of God is universal. Faith in an afterlife is nearly as universal. The nature of God and Heaven and Hell is a matter of faith. Heaven and Hell is best explained as a means to control human behavior in the form of a metaphoric carrot and stick. Before the formation of courts and prison it would seem that the invention of Heaven and Hell were considered necessary to control human behavior. But neither the threat of Hell nor the human means of controlling other humans by corporal punishment or imprisonment has proved effective in significantly changing destructive human behavior.

God is. Evil is. So is God by definition evil if God is omnipotent and omniscient? Could evil be created outside of God's power and knowledge? Evil is defined by Webster Merriam as something morally reprehensible. That is a very subjective definition as applied by and to human beings. Homosexuality is evil to religious fundamentalists all over. But the murder of a rape victim is not morally reprehensible to certain religious sects; in fact some consider it a moral imperative. Yet evil is part of everything. Are people evil or are there merely acts that are evil? Can a creation of God commit evil acts? Indiscriminate killing is universally evil. Are volcanoes and tsunamis which kill thousands innocent humans, indiscriminately, evil acts? Are acts of Nature which, according to Koran/Biblical fundamentalists totally in control of God, acts of evil perpetrated by God? I have described God here as perfect, defined as complete and encompassing all things. Evil is one of all those things. God cannot be complete, and therefore perfect, absent evil. God is everything which, by definition, must also encompasses evil. If evil is not part of God, God cannot be perfect i.e., complete. If God does not encompass evil then evil must be a thing which is out of the realm of God. But that is not possible because everything is God. God is.

Like God, Heaven and Hell actually exists or not. Philosophically Hell can be dismissed since a Creator God the Father cannot condemn any being He created if that being is fulfilling some essential niche in God's plan regardless of its abhorrent character.

Which leaves us to search for the existence of Heaven. Theists and animists alike share in the concept that every human possesses a soul which is the part of us that exists eternally. A soul being loosely defined as the spirit in a human being that lives on after the body ceases to function. Like God, if a soul exists it must consist of some sort of matter or energy. Is the soul a form of energy or matter? A photon? Quark? The question is rather does the soul live on as a separate entity possessing the essential qualities that existed in the human being but in an afterlife. Is there a separate specified part of the cosmos which is designated as heaven and/or hell where the soul exists or does the soul just float free? Does heaven or hell exist even philosophically? The problem is awareness. In order for a "soul" to benefit from the award of heaven or suffer the consequence of hell the "soul" must be aware of the award or consequence. Not even the most ardent fundamentalist can argue that the only way for something to be aware of its surroundings is through the brain. Is the soul constituted of specific brain cells, not necessarily in the same place leaving the body after death and somehow reforming in some other place into free floating brain un-supported by any known means? If a soul existed without the capability of awareness then the purpose of an afterlife would be moot. So once again the dilemma that the same God, Whom billions worship for providing the amazing natural laws which include the biology and physiology of the body, has to resort to magic in order to provide an afterlife for human beings.

In order for Heaven to exist for our benefit, then each of us must possess a soul. The human body has been examined to the smallest gene and virtually every molecule has been accounted for. That leaves the soul as a form of energy. Can a form of energy possess awareness? If the soul is merely a concept or idea or an emotion, then its existence is irrelevant unless a concept or emotion can be defined as a form of energy. But since energy can be reconstituted into matter, while an idea or concept fail to possess that quality, the existence of the soul has no scientific basis for existence. Insisting that the soul exists in a spiritual form is not an argument that can demonstrate its actual existence. Either the soul exists or it does not exist. Anything that actually exists must by definition exist in some form of energy or matter.

This same argument holds up against the existence of angels and Heaven. But if souls do not exist, then any argument supporting the existence of heaven is irrelevant since Heaven has no need to exist except to benefit a place for the soul. The argument might be made that Heaven exists in the entire Cosmos and does not exist as a specific place where souls are gathered. Still this does not provide a reason for its existence absent the actual existence of the soul. Furthermore, if Hell has no reason to exist then the existence of Heaven is equally superfluous. The only purpose for Heaven's existence is as an eternal refuge from Hell unless one totally redefines Heaven and its purpose.

If heaven actually exists in order for the human soul to have eternal life then the question that should be posed is: What is the reason for an eternal life? God by definition is infinite. But for what reason should any other form of life require eternal life? Hinduism's belief in continual reincarnation states a purpose for eternal life until one has reached nirvana but having reached nirvana the purpose for some existence is extinguished. If the purpose of God's plan for each human is to provide them a way to attain eternal life then He must have a purpose for rewarding eternal life.

Roman Catholic doctrine emphasizes the non-physical, spiritual, philosophical and metaphorical state of Heaven, and clearly places heaven in the category of the state of mind in oneness with God rather than a real place. Heaven is an idea, a feeling or a state of mind that lasts eternally in the soul. This eternal state of mind is utterly controlled by God Who is the judge as to whether our soul will be rewarded with eternal happiness or eternal suffering. But this argument then implies that a creation of God must have acted outside of His plan and therefore could be condemned for any acts that are committed although they were part of God's plan and done with his knowledge at the time of creation.

For all the religious martyrs (discounting suicide martyrs) there must exist those who killed the martyrs for their refusal to denounce their beliefs. The martyr goes to heaven confirmed as a saint. Spiritually martyrdom is a good thing. Absent the act of killing martyrdom does not exist. Therefore, the person or persons who killed the martyr are an essential part of God's plan and acting within His plan. If the death of a martyr is within God's plan then can God condemn those who carry out His plan? If anything happens for a reason it must happen within God's plan. Conversely if anything happens outside God's plan then it must happen without any reason and outside of God's plan which defines God as being either not omnipotent and/or not omniscient, and by definition not God. If God actually exists, God must be infinite, omnipotent and omniscient.

But if God is defined as the Creator, the question of His infinite actual existence must be questioned. God could not create that of which he is formed. If God is defined as the Father Almighty, the ultimate moral judge, that definition fails on every philosophical and logical level including His omniscience and omnipotence. If God actually exists He cannot be the Creator nor the Father Almighty and final judge. But, God must be infinite, omnipotent and omniscient.

5

Argument for the Actual Existence of God

THE ABSENCE OF ANY matter and energy equals nothing, a vacuum. Therefore if God actually exists, God must exist in some form of matter and energy. Therefore God only exists because matter and energy exists. Conversely God could not exist absent matter and/or energy. If God cannot exist without matter and energy can matter and energy exist without God? A theist must say no and an atheist must say yes. We know the law of conservation states that energy cannot be created nor destroyed and therefore must exist eternally by definition. God can actually exist if and only if matter and energy exists. If matter and energy may exist without God, is the converse possible? Could God have created that from which He Himself is formed? There are only two logical and physically possible explanations. God was formed separately and created from matter and energy; If God is a separate form of matter and energy then God the creator is a moot question since the same matter and energy which formed God could form everything that isn't God resulting in God as an unessential entity. The only possibility no matter how improbable it is: *God and all matter and energy are the same.* EVERYTHING IS GOD. This is Omnideism.

If God exists then Omnideism is the only rational, logical, and scientifically possible explanation for His existence.

It is argued that God exists spiritually, in a form that is neither energy nor matter but something nebulous and beyond human comprehension. This explanation surficially contradicts the theistic edict that "man was made in God's image". Does God exist as a factual concept like 1+1=2? Mathematical concepts such as this are more than idea because an idea must be generated. 1+1=2 whether it is generated as an idea or not. Pi would exist whether or not it was actually calculated. The fact of the aforementioned equation and Pi is not arguable. Does Pi or any provable mathematical equation actually exist? They are neither matter nor energy nor do they exist only because an idea was generated. The problem with existence as a concept, such as Pi or an equation, is that concepts are static and incapable of acting. A conceptual (spiritual) God would therefore be unable to perform the acts which are attributed to the Creator God and God the Father because a concept cannot act but can only be utilized for actions.

Omnideism syllogistically proposes, if God exists, then God must exist Omnideistically, and defines "exists" as a form that consists of matter and/or energy. By definition then a conceptual or spiritual God does not actually exist. So it could be posed more specifically, "If God exists as some form of matter and energy, then God can only exist Omnideistically."

But, if God is everything, then the argument of God as the creator is moot and only God's nature is at issue. If everything in the universe is a form of God then it may be arguable there exists some form of God which may exist as a separate consciousness. However, the possibility of some form of God as a separate consciousness presents the problem of God the Father and final judge which negates His omniscient and omnipotent nature which is essential to the defined nature of God.

Atheists, in many aspects, are as unreasonable as the most ardent fundamentalists and non-creationists. An explanation for the creation of the cosmos that does not include at least a possibility of God ignores the great enigma of the astrophysics. How do you get to the singularity preceding the Big Bang? The vast majority of physicists, astronomers, astrophysicists and the like all hold to the Big Bang theory of some kind or another. According to the most prevalent theory the universe, as we know it today and offered by Stephen Hawking, first appeared fourteen to fifteen billion years ago as a result of a tremendous explosion of energy which was, at that moment, just before the "Big Bang" contained in some kind of super black hole. The matter and energy that was emitted continues accelerating and expanding even as this is written. Atheists, who seem satisfied with this explanation of the beginning of the universe, blithely ignore the most obvious question, "Where did that super black hole or singularity come from?" The law of conservation concludes that energy and matter are interchangeable and therefore neither can be created or destroyed still leaves open the same question, From where did the energy or matter emanate? What started it all?

There is no physicist or astronomer who has responded reasonably and scientifically to that question with the exception of Roger Penrose and his theory of cyclical universes. It is an answer to which might be best proved, as Sherlock Holmes often stated, by eliminating all the impossibilities so then only the probability, no matter how unlikely, remains.

Absent scientific evidence to the contrary the only possible answer to the question is that matter and energy is infinite, consistent with the theory of conservation. If God actually exists then by definition God is infinite. And if God does actually exist then God must exist in some form of matter and energy and thus lays the more perplexing conundrum which constitutes the crux of this thesis.

Stephen Hawking's explanation of the origin of the universe that seems to include spontaneous combustion, a quantum theory of creation, contradicting accepted scientific theory. If energy and matter, or just energy did not exist prior to the singularity, there is no logical explanation, other than quantum spontaneity, as Hawking's clearly implies, as to how the black hole, that was the singularity, from which the universe originated and came to be. Then Hawking's theory cannot be argued without a magical rationalization and Hawking is left with the same conundrum as all atheists because he cannot logically find a beginning point. It is creation without a creator. Magic without the magician.

But, if all matter and energy is God, then there is no need seek a beginning point because by definition God is eternal and infinite as is matter and energy, which according to the laws of physics cannot be created or destroyed. And as matter and energy can be neither created nor destroyed then Hawking's explanation that nothing existed because time could not exist also contradicts the basic law of the conservation of energy. Hawking covers this problem by opining that the laws of nature only have existed since the beginning of time which didn't begin until the formation of the singularity and the Big Bang. The argument is embarrassingly circular. In contrast, Omnideism is consistent with both the laws of physics and the infinite nature of God. Thus, if matter and energy is infinite and God is infinite, they must be identical.

But what if all matter and energy is God then there is no need seek a beginning point, a moment of creation. A creation of any kind implies some unspecified being or thing that doesn't consist of matter and energy acted as a catalyst.

The question to consider is not the creation of the universe, but whether or not there is a rationally scientific explanation for the existence of the universe that includes God and conversely is there any rational explanation for the existence of the universe that cannot include God?

If God exists then there can be only one explanation consistent with the laws of physics and with the infinite nature of God as a supreme being. Matter and energy and God must be identical. If they are not identical we must deal with the chicken and egg conundrum. If God is a being, who actually exists as a separate entity from the rest of the Cosmos, God must still exist in some form of matter and energy and therefore matter and energy must by definition come before God. But if matter and energy came before God then God is not infinite which contradicts any acceptable definition of God. Furthermore, God cannot be the creator if

God exists as some form of matter and energy since God can't be composed of that which he had yet created.

Say God exists as a separate form of matter and energy then if matter and energy formed God there is no reason for God to exist as is the creator since matter and energy already contained the properties that can create all other forms of matter and energy. But if matter and energy existed prior to the formation of the singularity there is no logical or scientific explanation or theory for its formation other than creating itself. By definition God is infinite. By the law of conservation energy (and matter) is infinite. The remaining possibility, no matter how improbable is it that God and matter and energy are one. All matter and energy is God. Every atom, every quark, every photon is God. God exists as the single organism of the Cosmos.

This is the one plausible argument that is consistent with the actual existence of God and scientific theory, regardless of its unfathomability: God and matter/energy must be one and the same. And if they are one and the same then all the matter and energy that existed prior to the singularity that eventually formed the Cosmos at the time of the Big Bang is God. It is senseless to argue that God exists outside the matter and energy which formed the Big Bang because then one can only argue that God exists as something that is not matter and energy. However, if God and matter/energy are identical prior to and at the moment of the Big Bang then everything that exists, the Cosmos, is God. This is Omnideism. Everything is God.

If God exists, there can logically be no other alternative. God cannot be the creator of matter and energy because God is matter and energy. And since the laws of conservation preclude the creation of matter and energy there is but one conclusion. There is no creation and therefore no creator in the sense that we have traditionally defined creation and the creator.

If God is literally the energy/matter of the universe then the theology of such a statement is undeniable, in that now we can state, based on that premise, that God is not just part of everything in the universe but that God is EVERYTHING, not just in a theological sense but in the most basic physical sense. God is the computer I am typing on, the glasses I am wearing, the bacteria that covers my body, the carpet in my office, the fly that is buzzing outside the window, the sun, and the dark matter of space. So let's rewrite the beginning of Genesis this way:

> In the beginning all the matter and energy in the cosmos was compressed into a tiny ball, and that ball was God, and it was so compressed that it finally exploded into the nothingness creating all the elements of the universe according to His laws of nature. God, in the form of all the matter and energy in the universe eventually formed into the heavens and the Earth and an infinite number of other bodies speeding across and filling the empty universe including the sun and the stars which shone upon the Earth. As all the matter and energy of the Cosmos evolved into the simple plant and animal life forms which begat more complex life forms, including animals, male and female, evolving into a great variety of species some of which developed greater intelligence and understanding of their own beings one generation to the next until there evolved a multitude of intelligent animals. Some of these animals begat animals which could reason and understand and these animal forms of God are called Human Beings.
>
> And God's natural laws caused the rain and the snow to fall and the hail and caused the mountains to erupt and the earth to shake and the wind to blow all of which are God.

This "Genisis" is consistent with Roger Penrose's theory of a cyclical cosmos and that some form of matter and energy existed prior to the singularity, which formed this present universe, as a chaotic free-floating mass of sub-atomic particles such as quarks, positrons and photons randomly moving through space ultimately forming the singularity which preceded the Big Bang.

There is no other scientific explanation, theory or hypothesis for the appearance of the singularity or for what existed prior to the singularity. Omnideism hypothesizes a logical explanation for the existence of God. Now if a theory, or even a hypothesis, is introduced which logically and scientifically explains the creation of the singularity out of the chaotic distribution of energy, then Omnideism may be rendered a pointless intellectual exercise as most would render it now. However, absent that scientific theory, the formation of the singularity out of the chaos of energy (sub-sub-atomic particles: quarks, photons, and Higgs Bosun as well as the four forces.) is best logically explained because [matter and energy] = God. God is not important to the Penrose theory, but the actual existence of God as necessary for the formation of the singularity, is the only existing probability that is left after considering and dismissing all other possibilities as impossible.

It is argued by Hawking and others, that prior to the Big Bang, the laws of physics did not exist. Energy existed but without the formation of matter prior to the formation of the singularity causing the Big Bang. The singularity was the first instance, it can be hypothesized, that matter appeared as a result of the formation of the black hole singularity and consistent with natural law . Omnideistically, the formation of the singularity from the energy, the particle soup which existed pre-singularity, was the formation of God as a physical entity. The singularity, the ultimate black hole, the first combination of energy coalescing into matter, was God, Omnideistically. God in the form of the singularity, created all the laws of nature, which caused the Big Bang. God did not "create" matter and energy, because there is no creation. The formation of the singularity is the formation of God and the result of that formation, or the reformation for this aeon, according to Penrose, are the laws of nature, physics, chemistry, biology, etc., which resulted in the Big Bang and the formation of the Cosmos as it exists today which is the singular organism called God. Omnideistically each "aeon" is preceded by the formation of the singularity, which is God, and is necessary for the necessary Big Bang which forms another in an infinite number of Cosmos, one of which we are presently existing.

Further evidence that Penrose theory is more probable than not, might be found, in that, we know what will happen when the last star in the present universe burns out. We know the sun will burn out in another 5 billion years and probably become a giant red and consume all the planets. And eventually the sun will stop existing in any recognizable form and break apart into the sub-atomic particles from which it originally was formed. All the stars in the universe will meet the same end, and the Cosmos will only exist as a chaotic mass of energy—photons, neutrinos, and quarks. This is estimated to occur over the next 100 trillion years. This result brings us to the exact same circumstances that, theoretically, occurred prior to the Big Bang. If the four forces and the Higgs Boson existed, thus causing the singularity to form, there is no reason to think that the same thing couldn't happen again nor is it unreasonable to conclude that it happened before. A hundred trillion years is a nearly unfathomable time period except when compared to infinity. Penrose's theory that the Cosmos, as it exists today is just one of a succession of Cosmos formations which have resulted from the infinite nature of energy and the result of scientific principles is consistent with the Omnideistic infinite nature of the Cosmos, and the definition of God, and is a more probable explanation than the impossibility of spontaneous generation no matter how improbable the idea of infinity may be. The actual Omnideistic existence of God provides a further explanation of the formation of an infinite number of singularities resulting in an infinite number of Cosmos.

Omnideistically, if each aeon's Big Bang reconfigured the form of God as chaotic energy into the form of the Cosmos then a Euclidian analysis of Omnideism using well accepted definitions and postulates would result in the following proof:

- By definition energy and matter can be neither created nor destroyed, therefore energy and matter must be infinite in nature by the measure of both space and time;

- If matter and energy can be neither created nor destroyed then the quantity of matter and energy is inflexible and must only change in form;

- If, God exists, then by definition He is infinite in terms of space and time;

- If something actually exists it exists in some form of matter and/or energy;

- If God is a being of some kind and actually exists in some form then the being of God consists of, or is made up of, some form of matter and energy;

- There is no evidence of any other thing that exists in the cosmos that doesn't consist of some kind of or combination of matter and energy;

- Therefore, if God actually exists then His being must consist of some form of matter and energy;

- If God consists of matter and energy then, God could not have created matter and energy since God could not have created the very same thing from which His being is formed;

- God and matter and energy are both infinite by law and definition, respectively and both have existed simultaneously therefore there was no creation and no creator;

- (The Omnideistic God might better be defined as a "Generator" since the matter and energy needed only to be shaped into the forms of matter and energy we recognize as the Cosmos.

The more interesting question would concern the formation of the laws of nature which allowed or propelled the formation of the Cosmos prior to the formation of the singularity. Was the singularity formed as a result of the laws of physics? It is scientifically certain that, absent the laws of physics, the Cosmos could not recognizably exist. It is therefore a reasonable argument that the existence of the Omnideistic God created the laws of physics. If the laws of physics (or any other laws of science) infinitely existed, then would the Big Bang have necessarily occurred, as the Cosmos, in the form we know it, would have also infinitely existed, which we know is not true? But absent the laws of physics could the formation of the singularity have occurred without the subatomic particles, photons and Higgs Bosun, which would have been necessary to form the singularity. Penrose's theory does not fully explain how the form in which matter and energy existed, pre-singularity, could ultimately form the singularity? An Omnideistic God best explains the formation of the singularity consistent with laws of physics which directly caused each "aeon" to commence with a Big Bang.

The only Omnideistic alternative to Penrose's infinite cyclical Cosmos is that God, in the form of the singular organism that is the Cosmos, existed, in the infinity of space and time, in the form of the chaos of sub-atomic particles, prior to the formation of the singularity that concluded in the Big Bang and will continue to exist, in infinity, as a chaotic sub-atomic form, once the matter, i.e., stars and black holes, now existing in the Cosmos, has disintegrated. However, the concept that an Omnideistic God existed as sub-atomic chaos for an infinite time prior to the Big Bang, then in a finite time as the present Cosmos, and will then exist infinitely again as sub-atomic chaos, defines a very limited God. God's infinite nature would seem to be inconsistent with the very idea that God would only exist as a coherent single organism, we call the Cosmos, in one finite moment.

Absent Penrose's theory or the aforementioned infinite chaos, prior and pursuant to the formation of the singularity, or any other alternative theory that provides for the infinite nature of matter and energy, Omndeism is inapplicable. The infinite existence of matter and energy, consistent with the Law of Conservation, must be true in order to argue Omnideism as an explanation for the existence of God. However, if matter and energy is not infinite then God cannot actually exist.

If God is created from matter and energy then the laws of nature already existed and any creation from those laws is absent any input from God. If God can exist without being the creator then the argument against Omnideism is negated. Either way God's existence is then moot as a creator.

God can only exist if God and matter and energy are identical. And if matter and energy are infinite then, by definition, there can be no creation and no creator. If God exists then He can only exist Omnideistically.

OMNIDEISM IS NOT BASED UPON BELIEF. OMNIDEISTICALLY, GOD EITHER EXISTS OR DOES NOT EXIST

In order to prove the actual existence of God the following statements must be true:

1. Matter and energy can be neither created nor destroyed—law of conservation of matter and energy.

2. If matter and energy cannot be created or destroyed it must be infinite.

3. God by definition is infinite

4. Everything that exists in the cosmos, exists as matter and/or energy by definition.

5. If God exists He could not have been created, by definition.

6. If God exists then God exists in the cosmos.

7. If God exists then he must exist as some form of matter and energy.

8. If God exists as some form of matter and energy then God could not have created matter and energy.

9. If God is created from matter and energy then God is not infinite.

10. But since by definition God is infinite, God is not created from matter and energy.

11. If God is infinite and therefore not created, and matter and energy are infinite and cannot be created, then, if God exists then God and matter and energy are identical.

12. If God and matter and energy are identical, and infinite, then there is no creation and by definition no creator.

13. If there was a creation then God cannot exist.

14. If God exists then there can be no creation;

Corollaries to the above proof include the following:

A. God and the laws of nature are one and the same;

B. If the formation of the Cosmos was a singular event then God cannot exist;

C. If God exists then the formation of the Cosmos cannot be a singular event;

D. God is Everything that exists, has existed and will exist;

The Cosmos has various definitions depending on the dictionary but for Omnideism the Cosmos is defined as a single organism that is everything that exists. If Everything that exists, exists as the Cosmos then God is the Cosmos. God and the Cosmos are identical. God is the Cosmos. God is a single organism but Who exists in infinite forms.

Omnideistically, there is nothing metaphorical or spiritual in this statement. Nothing can exist outside of this single organism Which is God. Everything that exists is God. EVERYTHING. Every form of energy and matter. Every sub-atomic particle. Every photon. Every Higgs particle. It is all God. The only difference between a quark and a walrus is appearance and form of its Godness. A stone and a piece of lint are just different forms of God. Lightning and concrete are forms of God because everything that creates lightning and concrete is God.

This Omnideistic nature of God determines the omnipotent and omniscient nature of God. If God exists He is omniscient and omnipotent by definition. If something exists outside of the Cosmos then God cannot exist because He can no longer be defined as omniscient or omnipotent. God is infinite in space and time (and in all dimensions that include multi-verses, parallel universes and string theory). It is this infinite all-encompassing nature of God which determines His omniscience and omnipotence. If God is not everything then God cannot be omnipotent and omniscient. If a single atom, or positron or photon existed outside the organism, the Cosmos, which is God, then He cannot be defined as either omniscient or omnipotent since anything that form of matter or energy exists outside of the Cosmos, the single organism, is by definition outside of His control and knowledge is inconsistent with the definition of an omniscient and omnipotent God. If God is infinite in space and time then God is omniscient and omnipotent. God cannot be omniscient and omnipotent unless He is infinite in space and time.

6

The Cosmos as God

IT IS NECESSARY TO always keep in mind that the Omnideistic God is the dust on the floor, the heat from the sun and the sounds in the air. We have been brought up to believe that a being that is called God is all powerful and knowing and therefore anybody who proclaims themselves as God has those attributes. But that would defy the physical, biological, chemical laws of nature which restrict the potential growth of a human being much as they restrict that of an amoeba. It is only the degree of growth which separates humans from all other forms of God. God in human form has no more ability to defy scientific laws than does an amoeba. Furthermore if all humans are a form of God then no human is spiritually superior to another, except in their awareness of his or her Godness. Genetic abilities to attain a certain level of success within the restrictions of being human are merely part of the human form of the Omnideistic God.

What makes human beings special is that humans are the only beings, to our immediate knowledge, that have the ability to be aware of what they are, their form of being God, their Godness. Even those individuals who have a sincere belief that they can communicate directly with God are really doing nothing more than developing an insight into their own form of the Omnideistic being. Omnideistically, one could argue that those who testify to having direct communication with God are not aware that they are communicating with their own Godness rather than a separate consciousness.

Awareness that you are a form of God provides no tremendous physical advantage as Jesus demonstrated, miracles notwithstanding. The crucifixion and its aftermath expresses the indestructible and infinite nature of energy and matter as God.

If God is everything that makes up all the matter and energy of the universe can He be more? Can God be everything and an entity? Is God perfect? Yes, to all of these, if you use the Merriam Webster third definition of perfect which defines perfection as "pure, total, lacking in no essential detail, complete".

To think of God in this manner one has to divest oneself of any notion that God possesses a consciousness or intelligence that is analogous to a separate consciousness or in any way anthropomorphic. God was the singularity that preceded the Big Bang. All the energy (and/or matter) that created the singularity was God. Everything produced by the Big Bang is God. All dark matter, alpha rays, comets, planets, meteors, fungus, amoebae, and human beings, which have eventually been created following the Big Bang, are God, which resulted from the laws of physics, chemistry and biology that determined and continue to determine how God emerges in an infinite number of forms.

Most forms of God, inanimate objects and virtually all non-human forms of plants and animals, do not possess consciousness or intelligence. The consciousness and intelligence that we recognize as human is not a consciousness or intelligence that should be equated to an Omnideistic God's consciousness. God's being, consciousness, and intelligence, are the compilation of the Cosmos. There is no separate entity or consciousness that is God. The single organism of the Cosmos is God and everything that exists in that single organism is a form of God. Every thought and action of every animal is the thought and action of God.

I am going to try an analogy that might or might not work. Take a barrel of crude oil. We know that the oil in that barrel eventually takes the form of a plastic toy, a gallon of gasoline, a pair of nylon stockings, cosmetics, and a freeway. Imagine the barrel of oil as the singularity. (You also have to imagine that nothing needs to be added to the oil to form the various products.) Everything that is produced from that barrel of oil is still oil but in all its different forms.

God is the singularity (the barrel of oil) containing all the matter and energy (nothing added) that took form at the moment of the big bang.

Each individual form of God as has evolved through the same natural laws as the universe has. However, arguably, if there is another plane of reality, a parallel universe or multi-verses, they are all still God in regardless of the form the matter and energy has evolved.

Einstein was right, God does not play dice. Quantum mechanics are still a result of the Big Bang and consistent with the laws of physics whether humans totally understand the laws or not at this time. The infinite forms of God have evolved according to the laws of physics, chemistry, biology, mathematics and all the other sciences which are all God. The mere existence of scientific law is evidence that God exists Omnideistically. Stephen Hawking et al. have no explanation other that self-generation for the existence of natural law and revert to magic without a magician.

Is it possible that the infinite nature of energy, pursuant to the law of conservation, allows for the hypothesis that, prior to the formation of the singularity, energy existed in a chaotic formless state as quarks, but still, because the most basic laws of nature existed, enabled the chaotic energy to form the singularity? Consistent with this hypothesis, the Higgs Boson existed in this pre-singularity chaotic energy state, and then, in concert with the basic forces of weak, strong, electromagnetic and gravitational forces, formed the singularity from the existing energy, that, Omnideistically, was the only form of God, which preceded the Big Bang.

The coherent formation of the singularity from the chaos of energy can best be explained Omnideistically, where there is no magical spontaneous generation of energy as proposed by Hawking. Omnideistically then, God existed as the incoherent form of energy, quarks, Higgs Boson and the four forces, which formed the singularity fourteen billion years ago. The infinite existence of God, by definition, and energy, consistent with law of conservation, provides an improbable but possible explanation of the formation of the singularity.

If energy always existed, as the law of conservation states, then the most basic laws of physics must also be infinite in order to determine the formation of the singularity. If God doesn't exist could the incoherent energy and the laws of physics have existed and formed the singularity? It is possible, but why would the laws of physics suddenly form a singularity when the energy, the quarks and Higgs Boson, had already existed for an infinite amount? Omnideistically, the infinite existence of God as everything can explain the formation of the singularity, as a result of the effectuating laws of physics, (the Higgs Boson and the four forces) fourteen billion years ago. If God does not exist then it is simply more difficult to explain the formation of the singularity.

The infinite existence of energy is improbable but the only hypothesis which is scientifically possible. Spontaneous generation is scientifically impossible. Sans God, if energy, in the form of quarks and Higgs Boson, existed, and the basic laws of physics infinitely existed, then the formation of the singularity only fourteen billion years ago must be explained. But, Omnideistically, if God is everything then the eventual implementation of the basic laws of physics, which determined the formation of the singularity from the incoherent, but infinite existence of the quarks and Higgs Boson, is scientifically comprehensible.

Natural scientific law is consistent with the existence of the single organism of God, Who is the Cosmos. If energy and matter are interchangeable and only differentiated by their atomic structure at any moment then, as all energy and matter, God is indestructible and can be neither created nor destroyed. In this perception the theories of theology and science are no longer contradictory nor confrontational and in fact become a single entity. Omnideistically, theology and the laws of nature are interdependent as one cannot exist without the other.

7

God in Human Form

Everything is God.
Everything that exists is a form of God.
Every form of God is unique unto itself.

ONLY HUMAN BEINGS, WITHIN our very limited knowledge of what actually exists in the Cosmos, have the capacity to be aware of their own Godness and the Godness of those around them, and the Godness of the Cosmos.

Since there are billions of galaxies in the Cosmos and billions of stars in each galaxy and billions of planets revolving around each of those stars, the chances are very good that at least one of those other planets contains some form of intelligent human-like beings who also have the capacity to be aware of their Godness and the Godness of the Cosmos. But absent that knowledge no other form of God, has the capacity for that total awareness of the Cosmos. Awareness of one's Godness and the Godness of everything that exists in the Cosmos creates the responsibility for each individual to perfect one's own Godness and the Godness of the Cosmos. Obviously, the question is "How?"

Omnideism and Christianity absolutely agree that Jesus was a manifestation of God, except that Omnideism doesn't agree that

Jesus was unique because everything that exists is a manifestation of God. Omnideism and Christianity might also agree that Jesus perfected his Godness. Christians will argue that Jesus was perfect upon birth as "only" son of God, as opposed to successfully striving to perfect his own unique Godness. Jesus perfected his form of Godness through his actions and intentions to perfect the Godness of the Cosmos and the Godness of other human beings through his acts and teachings. Both Omnideists and Christians would agree Jesus lived without any possibility of receiving an after-life benefit because of his actions and teachings. Jesus, in Christian doctrine, acted without fear of eternal damnation nor in hope of receiving eternal happiness. No possible punishment, no possible reward. Omnideistically he spent his life striving to perfect the infinite Godness of the Cosmos.

This self-less and utterly altruistic approach is the essence of Christianity. Acting without seeking an award either in human form or in an after-life once that form ceases to exist as a human being. THIS IS WHAT JESUS WAS TEACHING!!! His parables are more consistent with Omnideism than with the Christian perception that every human should act in order to avoid hell or be rewarded with heaven.

The only way to perfect one's Godness is to act in a manner unique to one's form of Godness but to benefit the Godness of the Cosmos. But there is no one way to perfect one's Godness because every human being has a unique way of perfecting their Godness as a unique form of God. Perfecting one's Godness requires striving to fulfill one's ultimate potential, physically, emotionally, and intellectually, that are functions of, but within the limits of one's DNA and culture and circumstances.

Jesus was not in the position to be the world's greatest baseball player nor its greatest physicist. Conversely Babe Ruth and Albert Einstein were never in the position to symbolize a religious movement. And if they had been that would not have been a means of perfecting their Godness because it would not have utilized their talents to the ultimate. However, like the vast majority of human beings neither Babe Ruth nor Albert Einstein were the able to, or necessarily strived to, perfect all their Godness.

The epitome of Omnideism is knowing that there is no other reward except having the knowledge that you have done your best to perfect your Godness and the Godness of the Cosmos as Jesus did. Ghandi, Muhammad, Mary, Buddha, St. Theresa, and maybe Pope Francis the First, among countless others, for whom it could be reasonably argued that they perfected their Godness, without concern of a reward or punishment following death, because they were more aware of the responsibilities of perfecting their own Godness and the Godness of Cosmos.

Omnideism makes the very concept of a soul superfluous since all human beings, including their soul, would be considered merely a part of the human form of God. Furthermore, there is no need for heaven if the soul already exists as a form of God and will continue to exist just as God, the single organism of the Cosmos, continues to exist. The notion of Heaven implies that being a form of God is not the ultimate existence in and of itself. If God actually exists then the existence of heaven and a soul, Omnideistically, is a moot point for all forms of God, just as it was for Jesus, consistent with Christian doctrine. Nothing in Biblical literature gives the slightest hint or implication that Jesus had a soul. As the incarnation of God, having a soul would be pointless for Jesus and inconsistent with the Christian faith. God in any form doesn't need a soul to be saved. Omnideism, states 1) that any incarnation of God makes the concept of a soul moot, and 2) everything that exists is an incarnation of God, therefore the concept of a soul is purposeless.

Omnideistically the Christian symbolism of the Holy Trinity might be perceived this way: God the son represents all matter; God the Holy Ghost represents all energy; and God the Father represents the single organism of the Cosmos. Because energy and matter and the Cosmos are all the same thing by definition, just as traditional Christians believe the Father, Son and Holy Ghost are three in one, an Omnideistic Christian can demonstrate a three in one concept.

Muslims, Jews and Hindis could just as easily use Omnideistic principals to support their beliefs that the best way to perfect one's Godness is through their specific religious rituals.

The first thing I learned from the Baltimore Catechism and the Sisters of Mercy was that God knows everything, God is everywhere and God always was and always will be. These "mysteries" of God's infinite wisdom, infinite space, and infinite existence are explained as unfathomable in Christian doctrine. Omnideistically they are essential in order to rationally explain the actual existence of God. The last of these is clearly the exact same law of conservation that governs matter and energy. What we refer to as death is irrelevant. Most forms of God—stars, planets, dark matter, etc., do not live or die but merely exist. Most forms of God that are living—amoeba, bacteria, and dogs are not aware of their Godness. In our limited knowledge of the Cosmos, humans are unique in their capacity of having an awareness of their own Godness and the Godness of the Cosmos and therefore having the unique responsibility of perfecting their own Godness and the Godness of the Cosmos. Still, not all human beings bear that responsibility. Only those human beings that have the capacity for that awareness assume that responsibility. Children and the mentally incapacitated have no responsibility to perfect their Godness.

Whereas a Creator God, by definition, determines that there is a "plan" to which everything is subject and over which the Creator has total control, an Omnideistic God has no control because everything that exists is God and each form of God that has the capacity to recognize their own Godness has total control of their own actions and the free will to exercise their own Godness. A Creator God negates free will and possibility of humans to take responsibility for their actions since any act committed must be consistent with the Creator's plan. If any act were committed outside of the Creator's plan would mean that the Creator was not omnipotent and therefore cannot be defined as God. Only Omnideistically can humans be held responsible for acts that are culturally considered evil and detrimental to the perfecting of the Cosmos.

Omnideistically God is Hitler, Jesus Christ and a banana. God is Stalin and Pol Pot and all the murderers and rapists in the world. God is Abraham, Jesus, Mohammed and Ghandi and all the heroes and saints. God is a grain of sand, a computer program and the computer and the electrons and photons providing the energy for the computer. Only if God is all matter and energy and all its infinite forms can we rationally explain so many of the "mysteries" of Christian faith: explaining the holy trinity; Jesus both as the Son of God and the Son of Man; Jesus body as Holy Eucharist and wine as his blood. Only through Omnideism can we explain Muhammad and the preceding prophets as messengers of Allah and Hindi concepts of reincarnation.

All faiths dictate rituals and forms of worship that their followers must conform to in order to fulfill their obligations to the Creator God and which, within that structure, provide proof of one's faith and eventual entry into an infinite afterlife for humans. But such rituals and worship contradict the concept that the Creator God is omnipotent and omniscient and therefore in total control of everything that occurs in the Cosmos. The rituals and forms of worship are not acts which can perfect the Cosmos, the single organism, which is God. It was the rituals and forms of worship that Jesus argued against in his parable about the Publican and the Pharisee demonstrating that actions, not prayer or religious rites, were essential to perfecting one's Godness.

Religious ceremonies are nearly irrelevant, Omnideistically. Only when religious ceremonies, including prayer, services and sacramental rites are useful in encouraging followers to strive to perfect their Godness and the Godness of the Cosmos are they relevant Omnideistically. Only when religious rites help individuals recognize their own Godness, and the Godness of everything that exists in the Cosmos, are those rites relevant, Omnideistically. Omnideistically a Catholic, Baptist, Moslem or Hindi can all continue to claim that only by following their rituals can an individual truly recognize and perfect their own Godness and the Godness of the Cosmos. But clearly such claims are, ironically, a matter of faith, since every human has the capacity, regardless of any religious affiliation, or lack thereof, for recognizing and perfecting their own Godness, and there can be no proof or evidence that one set of religious ceremonies is more effective in striving to perfect one's Godness. Any prayer or worship to an Omnideistic God is anathema to Omnideism since there is no separate consciousness that exists outside of the single organism that is the Cosmos which is everything that exists.

Omnideism provides that one can, theoretically, know and understand the Omnideistic God, the single organism of the Cosmos, from knowing and understanding everything that is the Cosmos. In this manner science, philosophy, and the humanities are all equally able and necessary to lead to a better understanding of the Omnideistic God and one's own Godness and how to perfect that Godness and the Godness of the Cosmos. Pragmatically, since the Omnideistic God is infinite, it is impossible for any single form of God to know and understand the single organism of God Who is the Cosmos. But seeking to know and understand the Cosmos is essential in striving to perfect one's own Godness and the Godness of the Cosmos.

The form of the Omnideistic God as human beings has resulted in each human form of God evolving our own individual intelligence and consciousness. We are all connected individually and collectively in the Omnideistic Cosmos, each of us as a unique form of God, which has evolved from the singularity of matter and energy that preceded the big bang.

But if God is everything, Omnideistically, then those who believe that Jesus is a form of God needn't change their essential belief. Jesus is a form of God. Jesus is a form of God because Jesus existed. Muhammad is a messenger of God and a form of God because Muhammad existed. Buddha, Joseph Smith, and Martin Luther are all forms of God because they existed. The print on this page is a form of God.

The difference between these different forms of God isn't that Jesus is the only incarnation of God but, unlike most other human beings, Jesus' awareness of and understanding of his Godness and the Godness of the Cosmos was, arguably, superior to all other human forms of God, which have ever existed. Jesus had such depth of awareness and understanding of not only his own individual Godness but the Godness of everything that existed in the single organism of the Cosmos. The argument may certainly be made that some of the proclaimed prophets, including, Abraham, Abraham and Mohammed, had an awareness comparable to that of Jesus. Muslims can certainly claim that Mohammed in his awareness of his Godness and the Godness of the Cosmos was equal to Jesus. Jesus' Omnideistic awareness is more apparent because of his distaste of the religious ceremonies emphasized as necessary by the faithful. Omnideistically, Mohammed's teachings emphasize rituals of worship of a separate conscious entity as necessary to perfecting one's Godness, as opposed to Jesus' teachings which emphasized acting in such a way which would perfect the Godness of the Cosmos by recognizing the Godness of everything that exists in the Cosmos. Jesus' message to all other human beings was how to be the best form of God you are capable of being.

St. Francis of Assisi, Ghandi, Buddha, Mary, St. Theresa, Abraham Lincoln and Pope Francis might all be acknowledged as human forms of God that recognized their own Godness and the Godness of the Cosmos and strove to perfect that Godness with their own unique abilities and limitations.

What truly distinguishes Jesus is that he is the only one of the aforementioned, who is reported to have clearly inferred that he was the incarnation of God. It is of interest to note that Jesus at no time ever denied, explicitly or implicitly, that any other human being was an incarnation of God. Jesus' other significant remark in this theological area was reportedly to state that he was the Son of Man. There is, in that statement, Omnideistically, an inference that all mankind is therefore God. Omnideistically, if Jesus recognized that he was an incarnation of God, the Son of God, but also recognized that he was the Son of Man, the only way he could be both is if Man and God are identical.

THE UNIQUENESS OF HUMAN FORMS OF GOD

The depth of consciousness of the self as we understand the psychology of consciousness is unique to human consciousness. Consciousness and intelligence are only a result of the brain matter which we possess as human beings. Chimpanzees and other great apes have a shallower self-consciousness and self-awareness and all other animals possess a different and much shallower ability to formulate a consciousness and intelligence. Plants and rocks and other inanimate objects have no brain matter and therefore no consciousness or intelligence. Their matter and energy, Omnideistic Godness, cannot formulate ideas or feelings. The perfection of their Godness is a result of their existence. A rock or tree is Omnideistically perfect as it exists but has no ability or responsibility to perfect its Godness. The same is true of human beings that as a result of either a genetic disorder or medical condition is unable or no longer able to intellectually recognize or be aware of their own Godness or the Godness of anything else that exists.

If God is Everything then it is arguable that there is also a form of God which constitutes a separate consciousness that has knowledge of and control of all the other forms of God equivalent to a human consciousness which has knowledge and control of the human body. It could be argued that the consciousness of God exists as merely a form of energy as dark matter, a black hole, a worm hole, or plasma. But any form of an Omnideistic God with a separate consciousness contradicts an Omnideistic God's consciousness that already exists as the consciousness of all living forms of God. A separate consciousness of God results in the same conundrum that a Creator God causes. That some form of God is somehow in control of the Cosmos in which there is a "plan" to which everything is subject and eliminates both free will and responsibility of actions which Omnideism dismisses because this concept of an all controlling God produces an irrational manipulative God. A God we thanked for saving us from a terrible death resulting from the same brain tumor which he gave us.

Omnideistically, God can only exist as everything and God's consciousness is the collective consciousness of the single organism which exists as the Cosmos.

Biblical/Koranic argument that living is purposeless if there is no judgment of human action. Theists of all stripes hold the conviction that there must exist an entity whose values we are here to pursue. If human beings are a form of God then the purpose of human beings is to act in a manner consistent with the expectations of God's consciousness, the collective consciousness of the Cosmos. Human beings are the only form of God, of which we are presently aware, which possesses a consciousness that is capable of being aware of its Godness and the Godness of the Cosmos.

Morality is cosmic. Evil and Good are cultural interpretations of the morality of human actions. Morality is the totality of all the morality that exists in the Cosmos. Morality is the measure of one's actions in striving to perfect their Godness and the Godness of the Cosmos. The morality of any specific culture doesn't change the cosmic morality since the cosmic morality is collective of all forms of God. Since any human can theoretically, at birth, be taken from any one culture and inculcated with the values and morals practiced in another culture, values and morals cannot be genetic; cultural morality being very different from the cosmic morality.

Every culture has its own practiced morality which may be different from another culture's morality but it doesn't mean that either of those moralities are in conflict with the cosmic morality, Omnideistically, which defines the Godness of the Cosmos. Omnideistically the failure to strive to perfect one's Godness is the equivalent of the Judeo/Christian/Islamic concept of evil. Specific cultural acts of immorality in one society are often defined as acts of morality in another culture. The only question Omnideistically is whether that act can be conceived as perfecting the Godness of the actor and the Cosmos. Omnideistically, an immoral act is any act which is detrimental to or inhibits the ability of another human being to perfect their unique form of Godness and the Godness of the Cosmos.

Because Omnideism holds that every particle of matter and energy is God and it follows that the consciousness of the cosmos is a collective cosmic consciousness.

Omnideism doesn't take God out of Human society but, quite the contrary, puts God in the center of Human Society because God IS Human society. God is Man. Each human being holds within their form of God a "human" portion of that consciousness. Thus the moral values which we would like to consider universal are unique to our species and form of God. I should emphasize here that the existence of such values neither adds nor detracts from a human's Godness or any form of Godness. What makes humans unique forms of God is the ability to be aware of their Godness and the ability to strive to be better forms of God. Conversely, humans also have the ability to consciously fail to perfect their Godness and be detrimental to other humans' ability to perfect their Godness. No other form of God can be more or less than what it is. A rock is always a rock and a chimpanzee is always a chimpanzee without any awareness of its Godness or ability to perfect their Godness nor the Godness of the Cosmos. A human who fails to make the effort to become a better form of God is failing as a human and immoral because he/she freely fails to strive fulfill his/her Godness.

Biblically and Koranically it is argued, that there should be consequences, i.e., heaven or hell, for such a failure, or success, or there is no purpose for life. Does a rock, or amoeba, or even a chimpanzee have a purpose for life? Some fundamentalists would, and have argued, the purpose of all other forms of matter and energy are to serve humans. Omnideistically, since every form of matter and energy is a form of God such an argument is nonsensical. All forms of God exist and allow human forms of God to become more aware of his/her Godness and the Godness of the Cosmos as Jesus or Mohammed or Buddha were more aware of their Godness and the Godness of the Cosmos.

Why should there be consequences for failing to strive to perfect one's Godness? Since humans are forms of God the question becomes does God require consequences for failing to be the best form of God he/she can be? Should any form of God be punished? Can the single organism of the Cosmos which is God punish the specific form of God for failing to be aware of his Godness or the Godness of the Cosmos? Any consequence of Heaven and Hell for a form of God is nonsensical. It is the absolute obligation of each human, as a form of God, to strive to perfect and become more aware of their own Godness through their personal interactions with all other forms of God in the Cosmos. The consequence for failing to fulfill one's own form of Godness is the failure itself. Human's, to our limited knowledge of the Cosmos, are the only form of God which can fail to perfect their Godness or the Godness of the Cosmos, because no other form of God in the Cosmos, of which we are knowledgeable, has the awareness of their own Godness to be anything than the form of God they are. Only humans can be aware of their Godness and strive to perfect the Omnideistic God , which is of the Cosmos.

Biblical theists could argue that if God is Man then if Man wrote the Bible that makes the Bible infallible but Omnideistically all books would be infallible because by that argument God wrote all books. The nuance is that humans are a form of God and that form is only as infallible as that form is potentially able to be. No one form of God is perfect, if defining perfect as all encompassing, only the wholeness of the Cosmos, that single organism that is the Cosmos, is perfect, and that single organism is defined, by Omnideism, as God.

New Age philosophy states that the whole is greater than the sum of its parts and such is the basis of Omnideism. If all the pieces of the universe are God then the totality of God's various forms of existence is greater than the existence of all the forms of God in the Cosmos. The whole of the Cosmos is greater than the sum of all the parts of the Cosmos, the whole of God is greater than the sum of all the forms of God that make-up the single organism of the Cosmos.

8

Jesus, the Quintessential Messenger of Omnideism

As PREVIOUSLY STATED, WHAT separates Jesus from other human forms of God who were apparently aware of their own Godness and the Godness of the Cosmos is that Jesus is the only human who even inferred that he was God, incarnate. Abraham, Muhammad, Buddha, Mary, Ghandi et al., were never reported to have made any such claim. None of the major religions have ever taken issue with the historical accounts of Jesus, the human being, and that he ever acted in a manner inconsistent with what is reported in the Gospels or Epistles. Omnideistically, therefore, Jesus is the clearest example of a human form of God totally aware of his own Godness and the Godness of the Cosmos and, most importantly, acceptance of his responsibility to perfect his Godness and the Godness of the Cosmos. It is this total awareness of his own Godness that makes Jesus the ultimate example of how the human form of God should act Omnideistically. Again neither Abraham, Muhammad, nor any other human form of God has claimed to be God and this explicit failure to accept their own Godness eliminates them from this very particular point of discussion.

If one accepts the Christian (including Mormons) belief that Jesus is the only incarnate form of God who has ever existed, it must also be accepted that Jesus, despite his martyrdom and perfect human life, could not be rewarded for those acts on Earth with the eternal happiness of heaven. Just as certainly Jesus could not be threatened with an eternity in Hell. As God, Jesus, consistent with Christian doctrine was aware that he was God, and therefore, acted without any hope for a reward, nor fear of punishment, in an afterlife. However, if we view Jesus, Omnideistically, the message to all human beings is that it is the responsibility of all human forms of God is to perform acts on Earth without any consideration of, or intent to seek, reward or punishment in an afterlife, because the promise of an afterlife is irrelevant to a human's responsibility to perfect the Godness of the Cosmos.

Jesus' existence is more consistent with Omnideism, demonstrating, by example, how a human form of God should act without the hope of reward or fear of punishment but simply to perfect his form of Godness and the Godness of the Cosmos. Jesus' parables and actions indicate that he was equally aware of the Godness of all forms of matter and energy in the Cosmos. Viewed Omnideistically, Jesus was trying to teach that the reward for perfecting one's Godness is its own reward and it is more consistent with Jewish/Koranic/Christian beliefs that no human form of God should expect a greater reward for their acts than Abraham, Muhammed or Jesus, could have received.

There is nothing reported in the Gospels that explicitly or implicitly states that Jesus claims that he is the only incarnate form of God. His reported reference to himself as both the Son of God (by implication) and the Son of Man (explicitly 78 times reported in the Gospels) suggests, Omnideistically, that all humans are forms of God. Could Jesus be the son of Man and the son of God unless God and Man are the same? The alternative is that Jesus is the son of Man separately from the son of God. Therefore, Omnideistically, if Jesus is the son of God and Jesus is the son of Man (all human beings), then God and Man are identical, by the transitive property of equality. If God and man are the same then all men are incarnate forms of God, a form of God, who, is capable of recognizing their own Godness. As a form of Godness all human beings then have the responsibility to act as Jesus did always acting and striving to perfect their Godness and the Godness of the Cosmos.

If Jesus ever stated explicitly, or even by inference, that he was the *only* son of God or the *only* son of Man, certainly that would have been reported in the Gospels. The absence of such a report cannot be ignored. Such a statement or concept would have been simple for Jesus to convey either directly or by parable. But nowhere does Jesus isolate or disconnect himself from any other human being nor does he conversely separate other human beings from himself. The only attempt to differentiate and isolate Jesus, as an incarnate form of God, and therefore a superior form of God compared to other human beings, was done by other human beings following Jesus' death. But there is nothing in Jesus life, at least as reported in the Gospels, that suggests that this was his message. Viewed Omnideistically, Jesus' message was that each human form of God must recognize their own individual form of Godness, recognize the Godness of everything that exists in the Cosmos, and has the responsibility to strive to perfect their individual Godness and the Godness of the Cosmos.

9

Omnideism and Worship

OMNIDEISTICALLY, THE PURPOSE OF life is to perfect and fulfill your own Godness and the Godness of the Cosmos. Worship is only relevant where there is belief that there is a creation and an entity that acted as a creator because the faithful feel the necessity to demonstrate their loyalty and love for the creating entity to whom they owe all. But if one owes everything to a creator then the creator controls all that exists and there is therefore nothing that exists has a responsibility for any acts since those acts are, by definition, controlled by the creator. Worship is only necessary to show gratitude and love for an entity where there is faith that a creator exists. Worship is irrelevant where there is no creator. Omnideistically, God and matter and energy are infinite and there is neither creation nor a creator.

Every human is a form of God. Arguing that Omnideism eliminates the ability to personally relate with the Supreme Being and therefore eliminates the reason for human existence is diametrically opposed to the essence of Omnideism. Omnideistically the reason for existence is to perfect the cosmos which is a single organism that is God. Theists want to simplify God and focus on a being that exists as an enclosed entity and a separate consciousness that controls the Cosmos. Omnideistically God is infinite in time and space and matter and energy and all that exists is part of that single organism that is the Cosmos.

Omnideistically, every thought, action and verbalization is a communication with God and it is only our inability to recognize and be aware of all forms of Godness that interferes with that understanding that we are always communicating personally with God. Only human beings have the ability to be aware of the Omnideisitic God which exists in all matter and energy and is therefore the source of our growth as human forms of God. All thoughts are a direct communication with God. All communication is communication with God. Nothing exists outside of God.

Recognizing and acknowledging and being aware of the Godness in one's own-self is meaningless unless one also recognizes and acknowledges the Godness in all other forms in the Cosmos. Again only the human form of God, as far as we know in our limited knowledge of what exists in the Cosmos, has the capability of doing such a thing. And this ability makes it incumbent on each human form of God to reach their potential in their conscious awareness of their own Godness. Jesus, arguably, attained the ultimate level of Omnideism in perfecting his Godness and recognizing the Godness of everything that existed around him. Arguably Abraham, Mary, Muhammad, Buddha, St. Francis of Assisi, Pope John the XXIII, Ghandi, St. Theresa, and Martin Luther King perfected their Godness and recognized the Godness of the Cosmos. Was it their Omnideistic self-awareness which made them stand out from the rest of the world? What is more commonly called Holiness and piety, Omnideistically, might be interpreted as a greater awareness of not only their own Godness but, more importantly, the Godness of the Cosmos, the Godness of all forms of matter and energy.

10

Omnideism and Responsibility

AWARENESS OF ONE'S OWN Godness, without acceptance and acknowledgment of the Godness of everything that exists in the Cosmos, is a very dangerous thing because it leads one to believe that they alone, are God incarnate. Any number of dictators and who have attained a superficial awareness of their Godness but failed to ever recognize and/or refused to acknowledge the Godness of the Cosmos and Godness of all other forms of matter and energy that exist in the Cosmos lead them to believe that they alone in the Cosmos possess Godness and concluding that they themselves are the one and only God incarnate. Numerous kings, queens, czars, dictators and self-proclaimed religious leaders clearly fall into this category. King David (at least in his early life), Caligula, Alexander the Great, Henry VIII, Ivan the Terrible, Hitler, Stalin, and Idi Amin are the most obvious examples of human forms of God failing to recognize the Godness of all other forms of matter and energy. Omnideism requires the recognition and acceptance of the Godness of the Cosmos and everything that exists in the Cosmos.

While the failure to recognize the Godness of the Cosmos is extremely dangerous, because it provides motivation, not only to act without fear of consequence but with a false sense of absolute entitlement, the failure to recognize one's own Godness, allows one to act without responsibility, because any act can then be rationalized as subject to and the result of forces outside the control of the actor. This is the exact rationale that has provided moral cover for all those who believe in a creator God and which Omnideism rejects outright. Omnideistically no human form of God, who is capable of recognizing their own Godness, can rationalize their acts of cultural immorality, depravity or repugnant behavior by claiming that it is part of a plan or subject to the wish or whim of a separate entity. Omnideistically "the buck" stops with each individual's ability to make a decision as a form of God. Just as a "creator" God cannot share or pass on responsibility by blaming a more powerful being or force, neither can an Omnideistic human form of God blame any act they commit on a more powerful entity because each form of God is that entity.

Recognizing the Godness in only specific individuals, outside oneself, is less problematic and dangerous than believing that only you yourself possess Godness. Christians recognize the Godness of Jesus and as a singular incarnate form of God although there is nothing Jesus is ever reported to have said that suggests that he claimed he was the exception and unlike any other human being or any other form of matter and energy existing in the Cosmos. Jesus' reported reference to the lilies of the field in Matthew 6:25, is Omnideistically consistent with the Godness of all forms of matter and energy in the Cosmos and his recognition and acceptance of the Godness of everything that exists in the Cosmos.

Jesus' reported admonition to "Love thy neighbor as thyself" is highlighted in Christian doctrine. Omnideistically this "greatest of all commandments" is the bedrock of Omnideism. Recognizing and accepting the Godness of all human beings, the entire Cosmos, is the ultimate purpose of every human being because we are the only forms of God, which we know of, who can strive for such a goal. Achieving this goal, awareness and recognition of Godness in themselves and all things in the Cosmos, allows human beings to act consistently as an incarnate form of God.

Omnideistically, God's perfection is the completeness, including infinite space and time, which constitutes single organism that is the Cosmos. No singular form of God which exists as part of the Cosmos can be perfect because only the whole of the Cosmos can be perfect. Yet those forms of God which are incapable of recognizing their own Godness or the Godness of the Cosmos are perfect in the completeness of their form of Godness as it exists. A speck of dust, a drop of water, a star, a flower, a dolphin are all perfect forms of God in their singular limited form of existence because none of those forms of God can recognize and be aware of their own Godness and strive to perfect that form of Godness and the Godness of the Cosmos.

While human beings because of their capacity to recognize their own Godness and the Godness of the Cosmos we are still limited in our ability to understand how to perfect our own Godness or the Godness of the Cosmos. Human beings can only strive to perfect their own Godness and the Godness of the Cosmos within the limited capacity which is the essence of being a Human form of God. The Human history of war, mass murder, and destruction of the Earth is undeniable evidence that the ability of a human being to perfect the Cosmos is greatly restricted by our ability to understand the Godness of the Cosmos and how we can act to perfect our own individual Godness within the limited form in which we exist. What each human form of God must strive for is to act in a way that is intended to perfect the Godness of the Cosmos. The punishment or reward for the failure or success in perfecting one's Godness is found within each human form of God. God cannot punish or reward God. As a form of God the awareness of that Godness must provide its own reward or punishment. The punishment for failing to perfect one's Godness is failing to recognize the Godnesss of the Cosmos.

Any action taking that injures or forfeits the life of another form of God can only be justified if the individual taking the action can show that the action will perfect their own Godness and the Godness of the Cosmos. Self-defense or defense of another is certainly justified as protecting one's form of God. Capital punishment does not protect any form of Godness where life imprisonment is an alternative and therefore cannot be justified. Killing a form of God in order to provide sustenance to survive is justifiable Omnideistically i.e., a lion killing an antelope.

The ultimate question that must be answered by the human form of God is whether any action by a human can, at least arguably, be justified as perfecting the Cosmos, the single organism which is God.

11

Omnideism as an Umbrella

CHRISTIANS WHO FERVENTLY BELIEVE that Jesus is God incarnate, cannot be wrong, because so are we all God incarnate. Christians who espouse the energy of the Holy Spirit cannot be wrong since all the energy including dark matter in the Cosmos is God. Jews who believe that the Abraham spoke directly to God cannot be wrong because every human being speaks directly to God every time they speak. Muslims who believe that Muhammed was God's chosen prophet and spoke God's words cannot be wrong because all words, spoken, written or tweeted, are God's words. Hindis who believe that the cows are sacred animals cannot be wrong because the cows are also God. Similarly Hindis and all other believers in reincarnation cannot be wrong because God as all matter and energy is infinite and eternal and therefore any photon or quark that might contain the "spirit" of one being could certainly be contained in any other being's "spirit". Buddhism, Confucionism and Shintoism cannot be wrong because no matter what is the object of their worship that object is, Omnideistically, God. False Idol's do not and cannot exist Omnideistically.

Without worship diverse faiths look very similar. Acts of worship, i.e., prayer, and sacrifice are assorted rituals which are meant to be demonstrations of faith are the only things which divide religions. If God actually exists those forms of worship lose all their value and all that is left are the shared values of all religions which are beneficial to human existence.

Religious wars have been premised on the belief that a specific religion was superior to the others, and the other religion's belief would condemn those who held that "false" belief to be condemned to suffer in an afterlife and that it was (or is) the obligation of the "true" religion to save the non-believers from their own fate. Those differences are always s premised on a faith of God as the creator. A God, who is the ultimate judge. A God, who singularly and consciously controls life and death. A God, who, most significantly, is an anthropomorphic, separate conscious entity. The Omnideistic God is none of these. The Omnideistic God is all of us, and is only judged by our own acts toward all other forms of God and how those acts perfect the Godness of the Cosmos.

The consciousness of the Omnideistic God, which consists of the the collective infinite consciousness of the Cosmos is necessarily infinite in both time and space in that death as we understand it cannot interfere with or subtract from the consciousness of the Omnideistic God. The Omnideistic God is a living God constantly expanding, evolving and growing as the Cosmic consciousness is evolving and growing as each form of God, which has the capability, evolves and grows by perfecting its own form of Godness.

Furthermore, whether string theory or parallel universes and multiverses exist, the Omnideistic God is all of these and the consciousnesses that are part of these states of matter and energy are part of the Cosmic consciousness of the Omnideistic God which is a single organism consisting of the infinite forms of God.

How does Omnideism change a human perspective of God? Praying to God, asking for His forgiveness, or giving Him thanks for good fortune has no relevance Omnideistically. "God's will" has no meaning when everything that exists is God. Asking God to intervene is meaningless, because there is no separate conscious entity. Promises and threats of heaven and hell are not a consideration because the afterlife of Omnideism is the continual existence of the individual consciousness in the Cosmic consciousness and the limitless perfecting of the singular organism that is the Cosmos.

An Omnideistic God is far more compatible with the most important teachings of both Old and the New Testament in a great many aspects than is the Biblical/Koranic God. Jesus' most important lesson, "Love thy neighbor as thyself" takes on far greater significance from an Omnideistic viewpoint. On the other hand the first three commandments concerning, the worshipping of God on a particular day, or worshipping false Gods or taking God's name in vain are totally irrelevant. The moral demand to treat all humans equally and compassionately despite individual peccadillos becomes an imperative in an Omnideistic cosmos. Omnideistically, racism, sexism, etc., is a direct attack on the single organism which is the Cosmos that is God.

Omnideism explains the mystery of the Eucharist, as the body of Christ, better than the mystical theology of the Roman Catholic or Epicopalian sects of Christianity. The wafer is God. The wine is God. Of course so is the hot dog bun and the beer you had at the game. An Omnideistic God can only be as comprehensible as the entire Cosmos. Human beings have only explored a minute part of the Cosmos and therefore, by definition, have only explored a minute part of God. Understanding the Omnideistic God can be done only to the ability to understand the entire Cosmos. The intellectual entirety of humanity is just a minute fraction of the Cosmos, of the Omnideistic God.

Omnideism eliminates any rationale for the religious fundamental argument to exist. Omnideism equalizes all forms of worship whether they are polytheistic, monotheistic, or animistic. Omnideistically, there is no form of worship which is incorrect or blasphemous. Depending on how one counts or translates, Omnideism eliminates the first two commandments because it makes it impossible to worship a false God. And the elimination of the possibility of worshipping a false God eliminates one of the justifications for many wars.

12

The Meaning of Omnideism

IF GOD EXISTS, HE must exist Omnideistically. A non-creator, without a separate consciousness Who exists as all forms of matter and energy as the single organism of the Cosmos. Each form of God that has the ability to be aware of its own Godness and the Godness of the Cosmos has the personal responsibility, within the limits of their unique form of God, to perfect that Godness and the Godness of the Cosmos and everything that exists in the Cosmos. But what does it mean to perfect one's Godness?

The most effective way responding to that essential question is by responding to the following questions, which were originally asked and answered by Aquinas in his theses that attempted to explain God in terms of Christian doctrine. I have attempted to answer the same questions from an Omnideistic point of view.

I. OF GOD'S NATURE AND ATTRIBUTES

- *Who is God?*

God is everything. God is the Cosmos. God is all matter and energy that exists. Everything that exists is God.

- *What is meant by saying that God is a spirit?*

God occupies no specific body because God exists as all matter and God exists as all energy. Nothing exists as God in the form of a separate entity, distinguishable from any other form of matter and energy. The collective infinite consciousness of the Cosmos is God. The spirit of God is the collective consciousness of the Cosmos.

- *What does this imply in God?*
God is every being and every animate and inanimate object. God is every emotion and thought. Nothing exists that is not God. Everything that exists is a form of God.

- *Is God perfect?*
God is perfect in that He is everything.

- *Is God good?*
God is all that is good. God is all that is evil. Nothing exists that is NOT God

- *Is God infinite?*
Yes, God is infinite, for He has no limits (Aquinas theses VII. 1).

- *Is God everywhere?*
God is everything and therefore God is everywhere. God is the single organism we call the Cosmos.

- *Is God unchangeable?*
God exists as all matter and energy in all its forms regardless of how they change.

- *Is God eternal?*
If energy cannot be created or destroyed neither can God as they are the same.

- *Are there several Gods?*
No, there is only one God (Aquinas theses XI.).
God is everything. There cannot be another form of everything. If there is a parallel universe or multi-verses they also God in whatever forms they exist.

- *How do you prove that if God did not have these attributes He would no longer be God?*

God is perfect because he contains everything including all that is evil and all that is good. If God does not contain all evil then evil exists outside of God and outside of His control. But if evil is outside of God then God can no longer be defined as perfect since perfect requires that everything that exists in the Cosmos is God. God must be infinite because energy is infinite and if anything exists that is outside of God then God can no longer exist as God. God by definition is omnipotent and omniscient because God is everything. If anything exists outside of God then God cannot be omnipotent and omniscient and therefore cannot be defined as God.

- *Can we see God in this life?*

We can see God because everything is God. We cannot see anything that is not God. We cannot imagine anything that is not God.

- *Can we see God in heaven?*

Heaven does not exist because the existence of Heaven requires the existence of Hell. Since God is everything and everything is a form of God no form of God could be condemned to Hell and no form of God requires Heaven. God does not need a reward for existence and human beings are a form of God with a consciousness and intelligence and ability to recognize their own Godness and the Godness of the Cosmos. No other form of God, to our present knowledge, has the ability to recognize their own form of Godness and the Godness in all other existing forms of God.

- *How can we know God in this life?*

We know God by reason that God is. We know God by recognizing Godness in ourselves and recognizing Godness in all other existing forms of God. We know God by recognizing and accepting God is the Cosmos.

- *What is meant by knowing God in this life by our reason?*

All God's creatures and creations including non-creatures, i.e., rocks, trees, fish, amoeba, are a form of God. To recognize all forms of God in all that exists is perfecting human Godness. We perfect our Godness by being the best form of God we can be. Each human has the responsibility as a form of God which has the ability to recognize its own Godness to perfect that Godness and the Godness of the Cosmos. Each human is a unique form of Godness and perfects their Godness by perfecting their form of God through their curiosity, intelligence, physical attributes and emotional connections and by helping all other human forms of Godness to perfect their unique form of Godness. Only human beings because of their ability to recognize Godness in themselves and others have the potential to perfect their own Godness. All other forms of God, of which we have knowledge in the Cosmos, cannot recognize its own Godness and therefore are already perfect forms of God.

- *What is it to know Him in this life by faith?*

Faith, by definition, brings into question actual existence of God, therefore faith is irrelevant. Faith allows humans to question the. God either exists or does not exist. Recognizing the existence of God as the single organism that is Cosmos is all that is. If God does not exist then faith will not create his existence. Recognizing the Godness in all the different forms of God that exists in the Cosmos lets us know God. The acceptance that God actually exists eliminates faith. There is no place for faith in the existence of God.

- *When we speak of God, or endeavour to express our thought concerning Him, have the words we use a correct meaning?*

Words allow us to express our recognition of Godness in all forms and in ourselves and allow us to perfect our own Godness and the Godness of the Cosmos.

- *When applied to God and to creatures, have these word the same meaning or one wholly different?*

No other form of God in the Cosmos has the ability to recognize its own Godness and therefore cannot perfect its own Godness. A rock or a lion or a tree is already the perfect form of God because that is all they are capable of being. Human beings have the ability to perfect their Godness because of their ability to recognize Godness in themselves and the Godness in all things existing in the Cosmos. Perfecting our Godness is the ultimate goal and responsibility of a human being. Words allow us to perfect our Godness whether they are spoken or just thoughts.

- *Then whatever we may tell of God, and however exalted be our expressions concerning Him, for us God ever remains unutterable?*

When we communicate either with our own thoughts or with other people we are communicating with God because God is. All that exists is God and therefore all communication is a communication with God. Whether we listen to our own thoughts or the thoughts of others we are listening to God. When we talk to other creatures or inanimate objects we are talking to God.

III. OF THE DIVINE OPERATIONS

- *What is the life of God?*

The life of God is the life of the single organism, which exists as the Cosmos and the acknowledgement of Godness in all that exists in the Cosmos.

- *Does God know all things?*

Yes (Aquinas XIV. 5). Because God is all things, His knowledge is the collective knowledge of all forms of God existing in the Cosmos, which is a single organism.

- *Does God know our secret thoughts?*

Yes (Aquinas XIV. 10). Our thoughts are God's thoughts because we are a form of God.

Our thoughts need not be outwardly expressed by writing or verbalization in order to become part of the Cosmos because our thoughts restricted are not restricted by our physicality.

Because every person is a form of God any thoughts which are developed are a part of the Cosmos and part of God's collective knowledge and do not need to exist separately as spoken or written words in order to become part of the Cosmos and God's collective knowledge and wisdom. The Cosmos is a single organism existing as God and nothing escapes or can be kept apart from that organism regardless of the limits of its physicality.

- *Does God know the future?*

No.

- *How is this knowledge in God?*

God is infinite. God is omniscient. God's knowledge is the collective knowledge of all forms of God that have knowledge in the Cosmos, which cannot know or control the knowledge of the future even as each form of God that has a consciousness has the free will to act as its own unique form of God. As the knowledge accumulates so does perfecting the Godness of the Cosmos. It is the human form of God's responsibility to perfect the Godness of the Cosmos by perfecting our own Godness. There is no separate consciousness that exists as a separate entity, that is God, that can know the future. God cannot predict God's future. If God knows the future then there could be no free will.

- *Has God a will?*

No. God is.

- *Do all things depend on the will of God?*

No. God does not have a will that exists outside the collective consciousness of the single organism of the Cosmos. Only the collective will of those forms of God, such as humans, which have individual wills and consciousness, that together form the will and consciousness of the Cosmos, the single organism which is God.

- *Does God love all His creatures?*

All forms of matter and energy are forms of God which exist as a single organism of the Cosmos. The love of God is the love of the single organism which is the Cosmos which is all matter and energy.

- *Does God's love for His creatures produce any effect in them?*

Humans are capable of love because only humans on Earth are capable of recognizing the Godness of all forms of matter and energy in the Cosmos. Love is a form of recognizing the Godness of all forms of God in the Cosmos and striving to perfect the Godness of one's self and the Godness of all other forms of God is a demonstration of love. The effect of one's love is how much the acts to perfect the Godness of the Cosmos by striving to perfect the Godness of all other forms of God. Any form of God that is capable of recognizing its own Godness has and the Godness of the Cosmos has the capability to love. The collective love of these forms of God is God's love.

- *Is God just?*
  God is everything.

- *Why is God Very Justice?*
  Omnideistically, everything is God and each form of God, is. The justice of God is the result of each form of God acting consistently to that form of Godness. Only human beings, to our knowledge, can perfect their form of Godness. The individual who acts for justice is perfecting their Godness and the Godness of the Cosmos.

- *Is there any special kind of God's justice towards men?*
  Perfecting one's Godness and the Godness of the Cosmos is award enough. As a form of God human's perfect their Godness within their own unique limits and as a form of God receive no special reward outside of their knowledge that they have acted to perfect their Godness.

- *Does God reward the good and punish the wicked in this life?*
  God, is not a separate consciousness or entity in the Cosmos but rather God exists as the entirety of the Cosmos, as a single organism. There is no punishment or reward from God, but only the absence of perfecting one's Godness and the Godness of the Cosmos. Human forms of God mete out punishment to other human forms when those humans act in a way that is contrary to the act of perfecting one's Godness or is detrimental to the Godness of the Cosmos.

- *Where does God fully reward the good and punish the wicked?*

The knowledge that one has perfected or failed to perfect their own Godness and the Godness of the Cosmos is the only reward or punishment.

- *Is God merciful?*

No. Mercy is not needed or relevant to the existence of God. God is everything and the forms of God that are wicked are punished or rewarded as they exist in human form for their actions detrimental to the perfection of the Godness of the Cosmos. Being a form of God in itself does not justify reward or punishment. There can be no reward or punishment for being a form of God other than the failure to perfect one's Godness once existence in human form ends.

- *Has God any care of the world?*

God is a single organism we call the Cosmos and exists as it is.

- *Does the providence of God extend to all things?*

The natural laws of science are a form of God and part of the single organism of the Cosmos and only by understanding those natural laws can we know the future and make predictions which are part of the collective knowledge of the Cosmos which is God. Providence as a form of faith in God's plan is irrelevant and cannot exist if God actually exists.

- *Does it extend also to inanimate things?*

God is all natural disasters and is all forms of evil and wickedness and pain and suffering and a result of the laws of science which are a form of the Cosmos, the single organism which constitutes the whole of God.

- *Does it extend to the free acts of man?*

Humans act within the limits their physical and intellectual forms of God, and are free to act in order to perfect their Godness or not perfect their Godness or to act detrimentally to the Godness of the Cosmos.

- *What does predestination imply with regard to those whom it concerns?*

As there is no creation there can be no predestination.

- *What are those called who never attain to this happiness?*

Eternal happiness is the perfection of the single organism that exists as God which is the Godness of the Cosmos.

There is no "heaven" which exists in the Cosmos but only the eternal knowledge that exists in the single organism that exists as the Cosmos which is God. Forms of God, such as human beings, recognize their own Godness and have made their best effort to perfect their Godness and the Godness of the Cosmos. Each human form of God has the obligation to perfect their lives and is responsible to help perfect the lives of all human forms existing around him/her perfecting the Godness of the Cosmos. There is no reward of heaven, because perfecting one's Godness is the reward of each human being who is a form of God.

- *But those who do not reach heaven, will they be punished for not getting there?*

God is a single organism which is the Cosmos existing as all matter and energy in all its infinite forms in the Cosmos including human form and a human form of God cannot be punished or rewarded since God cannot be punished for acts that are God's acts.

- *Do those who respond to God's offer and who reach the happiness of heaven owe it to God that they responded to His offer, and is it due to Him that they merit their happiness?*

Since there is no creation, and therefore no Creator, there can be no predestination because everything is God.

- *What does this choice imply with regard to those whom it concerns?*

Human beings have a unique capacity on Earth to perfect their Godness because only humans can be aware of their Godness. The purpose for each human being is to perfect their own Godness in order to perfect the Godness of the Cosmos. Humans are the only form of God that can choose to be detrimental to the perfection of the Godness of the Cosmos by hindering other forms of God from perfecting their own form of God and therefore the Godness of the Cosmos.

Humans must first recognize their own Godness then recognize the Godness in all other forms of God that exist in the Cosmos and strive to perfect not only their own Godness but the Godness of the Cosmos.

- *Is God almighty?*

No. God is. God is not a separate consciousness or entity that makes decisions or plans

All that exists is God. The Cosmos is God. God cannot be separated from the Cosmos and conversely the Cosmos cannot be separated from God. The Cosmos and God are indistinguishable.

- *Is God happy?*

God's consciousness is the collective consciousness of the Cosmos which is a single organism. God is all that is happy and all that is sad. God is all that is angry and all that is satisfied.

IV. OF THE DIVINE PERSONS

- *What is meant by saying that God is a spirit in three persons?*

God is everything. The entire Cosmos has the attributes of Godness but only human beings, to our present finite knowledge, possess the capacity to recognize and perfect their own Godness and the Godness of the Cosmos.

- *Who is God the Father?*

There is no God the Father because everything in the Cosmos is a form of God and no form of God is superior to another in its form of Godness.

- *Who is God the Son?*

All human beings are incarnate forms of God within the single organism of the Cosmos. Jesus was no different from any other form of God except in his success in striving to perfect his form of Godness and the Godness of the Cosmos. The same could be said of Muhammad, Abraham, Mary, Buddha and Ghandi because all demonstrated to all other Humans how to perfect their own unique form of Godness by recognizing and perfecting the Godness of all forms of Godness in the Cosmos. Every human being is an incarnate form of God and has the capability of perfecting their unique form of Godness.

- *Who is God the Holy Ghost?*
  There is no Holy Ghost. All the energy that exists in the Cosmos is God.

- *Are these three divine persons distinct from God Himself?*
  No.

- *Are they distinct from each other?*
  All forms of God are unique.

- *What is understood by saying that the divine persons are distinct from each other?*
  Each form of Godness is unique in the Cosmos as part of the Godness of the Cosmos. Only the human form of Godness can strive to perfect their Godness and the Godness of the Cosmos within the limits of their unique form of Godness.

- *Can these three persons be separated from each other?*
  All forms of God are a unique part of the Godness of the Cosmos which is a single organism that is God.

- *Were they together from all eternity?*
  The Cosmos and God, existing as the identical form of matter and energy, cannot be created or destroyed and are, therefore, infinite in existence.

- *Has the Father, in relation to the Son, all that we have affirmed of God?*
  God is the single organism we call the Cosmos.

- *And have the Son in relation to the Father, and the Father and Son in relation to the Holy Ghost, all that we have affirmed of God?*

God is the Cosmos and the Cosmos is God. Everything that exists, exists as a single organism that is God.

- *Are these three, thus related to each other from all eternity, three Gods?*

There is one God. God is the single organism called the Cosmos. Everything that exists in the Cosmos is God. But no form of God exists as an entity in and of itself outside of or apart from the Cosmos which is the single organism which is God.

- *Do these three persons form a veritable society?*

The Cosmos is a perfect society, in the sense that the Cosmos is includes everything that exists. The Cosmos is God and the Cosmos consists of all matter and energy which exists and each form is a unique form of God and collectively these forms of God is the Cosmos.

- *Why is this society the most perfect of all societies?*

Because the whole of the Cosmos is God.

- *How do we know there are three persons in God?*

There are an infinite number of forms of God because everything is God.

- *Could reason, without the help of faith, know that there are three persons in God?*

Faith is irrelevant. God either exists or does not exist.

- *When faith tells us there are three persons in God, can reason understand this?*

Faith is irrelevant. God either exists or does not exist. If God exists he can only exist Omnideistically as a single organism. Faith and belief does not affect the actual existence of God.

What are these truths called that are beyond reason's grasp and are known by faith only?

There are no mysteries. All truths which cannot be explained are a result of our inability to comprehend the explanation.

- *Shall we ever come to know the Holy Trinity in itself?*

There is no trinity there is only the infinite space/time/matter/energy nature of God. Omnideistically the Holy Trinity, may be compared in the following way: God the Father is equivalent to the wholeness of the Cosmos; God the Son is equivalent to all forms of God that have the capacity of recognizing its own Godness (to our limited knowledge only Human Beings fill this category); the Holy spirit is equivalent to all the energy that exists that is not formed into matter. The spirit, or energy, is everywhere and exists in everything.

- *Is it possible on earth to get a glimpse of the beauties of this mystery of the Holy Trinity by a consideration of those actions which are proper to intellectual beings?*

Only those forms of God which have the intellectual capacity to recognize their own Godness and the Godness of all that exists in the Cosmos have the responsibility to strive to perfect their unique form of Godness

- *Is there any kind of order among the Divine Persons?*

Every person is a form of God and equal in their capacity to perfect their Godness if they have the capacity to recognize their own Godness. No form of God is superior to another but they are unique and equal in the perfection of the Cosmos.

- *When the Divine Persons produce acts other than those known as notional acts (which are the acts of generation and spiration), are these acts produced by the three persons in common?*

There are no acts outside God because everything that exists is a form of God and all acts are a result of natural laws which emanate from the existence of all matter and energy which exists as infinite forms of God.

- *But are there not certain actions or certain sources of action which are attributed more particularly to this or that person?*

Any form of God which lacks the intellectual or psychological capacity to recognize their own Godness, and therefore cannot recognize the Godness in the Cosmos, no longer retains the responsibility of perfecting their Godness.

- *When therefore we speak of God in His relation to the world, do we always imply that it is God as one in nature and as three in person that acts?*

God and all matter and energy are identical. Everything that emanates from matter and energy is God.

V. OF THE CREATION

- *What is meant by saying that God is the Creator of all things?*

If God exists then God is infinite by definition. Matter and Energy cannot be created or destroyed and is therefore infinite by the law of conservation. If God exists then God exists as matter and energy. Therefore there can be no creation since something that is infinite cannot be created or destroyed. God and matter and energy are indivisible. There is no form of matter and energy which is not a form of God. God could not create that of which He is made.

- *There was then nothing at all beside God before He made all things?*

God is all things and could not make Himself.

- *When did God thus make all things out of nothing?*

Matter and Energy cannot be created or destroyed because by natural law of conservation all matter and energy is God.

- *Had He so wished then, He need not have created the things He has made?*

All forms of matter and energy are the result of the laws of nature which are God. There is no separate consciousness which controls the laws of God Who is everything and which we call the Cosmos

- *Why therefore did God wish to create at some given moment the things He has made?*

There is no creation. God is everything that exists and has ever existed and will exist in the single organism which is the Cosmos.

- *It was not then through need, nor in order to acquire some perfection, that God created the things that He has made?*

Everything that exists is a form of God and does not require further perfection except for those unique forms of God who have the capacity to recognize their own Godness and the capacity to perfect that Godness and the Godness of the Cosmos.

- *Do the angels exist somewhere?*

No. The existence of angels has no rational explanation nor reason to exist as everything that exists is God.

XIII. OF MAN: HIS NATURE; HIS SPIRITUAL AND IMMORTAL SOUL

- *Is there anything in this world which forms as it were a world apart, a being that is wholly distinct from the rest of the world created by God?*

To our limited knowledge a human being is the only form of God in the Cosmos that has the capacity to recognize its own Godness and the Godness of all things that exist in the Cosmos and therefore has the responsibility to strive to perfect its Godness and the Godness of the Cosmos.

- *What is man?*

Man (sic) is a form of God as is everything in the Cosmos, but is unique, in its limited knowledge of the Cosmos, in his capacity to recognize not only his own Godness but the Godness of all other forms of matter and energy.

- *What is the spirit called that is in man?*

The spirit in a human being is each human being's knowledge and recognition of their own Godness. Those human beings that don't recognize their own Godness include children and the intellectually impaired who have lost or never developed that capacity. The Godness of those who are intellectually impaired removes from them the responsibility of perfecting their Godness and the Godness they possess has reached its ultimate level as has the

Godness of any non-human form of God in the Cosmos which does not, by its own nature have that capacity.

- *Is man the only being in the world of bodies that has a soul?*

Only humans, on Earth, have the capacity to recognize their own Godness. There are, likely, other forms of Godness in the Cosmos who also have the capacity to recognize their own Godness but of which human's are not aware.

- *What is the difference between the soul of man and the souls of plants and animals?*

No plant, animal or other forms of God have the ability to recognize its own Godness and the Godness of all other forms of God in the Cosmos and only does a human being have the responsibility to perfect their own Godness and the Godness of the Cosmos.

- *Is it then by intellective life that man is distinct from all other living beings in this world?*

Yes.

- *Is this intellective life of the soul of man, in itself, independent of his body?*

Every human's capacity to perfect their own Godness is unique to their physical form of God. But the recognition of all other forms of Godness in the Cosmos is independent of a human form of God. A human's intellectual capacity to strive to perfect their Godness and the Godness of the Cosmos is unique within our limited knowledge of the Cosmos.

- *Can any reason be given to establish this truth?*

Absent the development of a Human's unique ability to recognize their own Godness and the Godness of the Cosmos all other forms of Godness there is no ability or responsibility to perfect their Godness.

- *But how does it follow from this that the human soul in its intellective life is, in itself, independent of body?*

Any form of God which has the responsibility to perfect its own Godness must not look for a reward other than the perfection of the Godness of the Cosmos.

- *What follows from this truth?*

God and the Cosmos are the same and cannot be created nor destroyed so that the Godness of each human is infinite. Human beings, as a form of God, must strive to perfect their individual Godness, the Godness of every other human being and the Godness of all that exists in the Cosmos, which is the single organism which is God.

The perfection of the Godness of the Cosmos is dependent on each form of God, which has the capacity to be aware of their Godness , that that each form of Godness that has the capacity, strive to perfect their Godness and the Godness of the Cosmos because the Cosmos and God are the same. The perfection of the Cosmos is the encompassing of everything that exists in the Cosmos.

XV. OF THE MIND AND ITS ACT OF UNDERSTANDING

- *Are there any other powers of knowing in man?*

Humans, in our limited knowledge, are the only forms of God with a capacity to recognize its own Godness and the Godness of all other forms of God in the Cosmos.

- *What is this chief power of knowing in man called?*

Intellect and reason

- *Is reason and intellect one and the same power of knowing in man?*

Yes.

- *Is reasoning an act proper to man?*

Humans, alone on the Earth, have the capacity to recognize their own Godness and the Godness of all other forms of God in the Cosmos. Those humans who recognize only their own

Godness will fail in striving to perfect their Godness or the Godness of the Cosmos. Those human's who lack the ability to reason lack the ability to recognize their own Godness or the Godness of the Cosmos.

- *Is it a perfection in man to be able to reason?*

Humans must strive to perfect their Godness although no Human being can ever actually perfect their Godness because their perfection can only be a part of the perfection of the Godness of Cosmos which is always being perfected and is the collective Godness of all that exists in the Cosmos.

- *Why is it a perfection in man to be able to reason?*

Humans, who have the capacity to reason and recognize their own Godness share responsibility to perfect their Godness but can only strive to perfect their Godness because the Human form of God is only a part of the Godness of the Cosmos which is God in its totality.

- *Must we say that man was made to God's own image and likeness?*

Humans are a form of God like all forms of matter and energy in the Cosmos.

- *What is understood by this?*

Everything that exists in the Cosmos is there to perfect the Godness of the Cosmos through its existence.

- *Is it possible to show how the nature of man and his actions, viewed in their highest endeavour, enable him to know God in His spiritual nature and to catch a glimpse even of the intimate life of the three Divine Persons?*

A human's highest endeavor is striving to perform to the best of one's ability limited only by their physical and intellectual capacity. The collective Godness of each form of God is creates the Godness of the Cosmos.

- *How can we imitate the perfection proper to the Divine Persons?*

There are many examples of humans who successfully strove to perfect their Godness including, Jesus, Muhammad, Buddha, Confucius, Mary, Ghandi, Mother Theresa and Abraham. Each of these humans showed the rest of humanity the actions which are required to perfect one's Godness. They each recognized their own Godness and the Godness of all forms of matter and energy in the Cosmos. Their actions interacting with the Cosmos were positive acts in perfecting the Godness of the Cosmos. No form of God in the Cosmos can be perfect except the whole of the Cosmos which is the single organism, God.

- *In the corporeal world is man only made to the image and likeness of God?*
  Everything is a form of God.

XVIII. OF THE STATE OF HAPPINESS IN WHICH MAN WAS CREATED

- *Was man created by God in a state of great perfection?*
  There is no creation or creator. Humans, until they have gained the intellectual capacity to recognize their own Godness, do not have the responsibility to perfect their unique form of Godness, however once a human has the formed the capacity to recognize their Godness they are responsible to strive to perfect their unique form of Godness and the Godness of the Cosmos.

- *What did the state of perfection in which man was created comprise?*
  The perfection of the God is the collective perfection of the infinite forms of God which create the single organism that is the Cosmos.

Some human's may not develop or may lose their capacity to recognize their Godness and thus do not have at that moment the responsibility to strive to perfect their Godness or the Godness of the Cosmos but such a failure does not lessen their own Godness. The perfection of one's Godness is measured by the actions which perfect the Godness of the Cosmos. Acts of evil do not detract from or negate perfecting one's Godness because acts of evil are necessarily part of the perfection of the Cosmos. However, acts which are deliberately detrimental to the ability to perfect one's Godness are evil because they are detrimental to the perfection of the Godness of the Cosmos.

- *Was man created by God in a state of happiness?*

Human's are born as perfect forms of Godness without need to further perfect their Godness until they become intellectually aware of their own Godness and the Godness of all matter and energy existing in the Cosmos and at that moment become responsible for perfecting their own Godness and the Godness of the Cosmos.

- *Was this state that of his final and perfect happiness?*

Perfect happiness is found only by recognizing one's Godness and the Godness of all matter and energy that exist in the Cosmos and then perfecting one's Godness and the Godness of the Cosmos

XX. OF GOD'S ACTION IN THE GOVERNMENT OF THE UNIVERSE; AND OF MIRACLES

- *How does God govern this universe which was created by Him?*

There is no separate consciousness or being that governs the universe or Cosmos. Each form of God that has the capacity to be aware that they are a form of God and recognizes their Godness is solely responsible for any act they commit physically or intellectually. The Cosmos is subject only to the natural laws of science which are part of the Godness of the Cosmos.

- *Does God maintain all created things Himself?*

All forms of God are the result of the natural laws which formed the Cosmos which exists as God in its perfect totality.

- *What is meant by saying that He Himself maintains all created things?*

All things existing in the Cosmos exist as a form of God. The totality of the Cosmos is a single organism that is God.

- *Is the act of maintaining all things in existence also proper to God as is the creation of things?*

The totality of the Cosmos is greater than the sum of the forms of God which form the Cosmos because only the totality of the Cosmos is God.

- *Could God effect that all things that are should cease to be?*

God is all matter and energy and is infinite by definition and natural law. Things that make up matter and energy, quarks, neutrinos, photons, et al., do not cease to be but can only change the form of their Godness.

- *What action on God's part would be necessary to effect that all things that are should cease to be?*

Energy cannot be created nor destroyed but only the infinite forms of God which exists as the Cosmos can be changed to other forms. The Godness of each form of God is infinite and can only take a different form.

- *Without ceasing, therefore, the being of all things that are in the world depends absolutely upon God?*

The Cosmos in its totality is God and each form of God is dependent on all other forms of God existing in the single organism which is the Cosmos.

- *Why has God never annihilated and will never annihilate anything He has created?* ~

God is infinite. Matter and Energy are infinite. God is all matter and energy and cannot be created or destroyed. The Cosmos is God. God is the Cosmos. Nothing that exists is NOT God.

- *Can there be any change in things made by God?*

Those forms of God, such as human beings, which have the capacity to recognize their own Godness have the nature and responsibility to perfect their own form of Godness have the capacity to perfect the Godness of the Cosmos.

- *Do these changes which sometimes come about in the things made by God; enter into the plan of His divine government?*

There is no plan or government that is controls the acts of any form of God outside of natural science or the intellect of a form of God that recognizes its Godness.

- *Is this action which is proper to God and to which we must attribute the changes that come about in material things "outside" the action of secondary causes which in the ordinary course of nature is proportionate to these changes?*

Everything that happens in the Cosmos is a result of the natural laws of science which are the result of forms of matter and energy which are God.

- *Are there any such miracles performed by God?*

The totality of the Cosmos is God and everything that occurs in the Cosmos flows from the natural laws of science of the Cosmos. Miracles are acts or circumstances which occur within the laws of science and nature but which at the time they occur are outside the ability of humans to understand.

XXI. OF THE ACTION THIS GOVERNMENT; OF THE UNIVERSE OF CREATURES IN AND OF THE ORDER OF THE UNIVERSE

- *As regards the changes that come about, or can come about in created things, can creatures act and do they act one upon the other?*

The actions of different forms of God are unique to each form of God and the result of that form of Godness. Each act by any form of God results in the perfection of the Cosmos. The Godness of the Cosmos is perfect as a result of its totality of all forms of God. All acts of any form of God perfect the Godness of the Cosmos except those acts which deliberately interfere with another form of God's ability to perfect the Cosmos.

- *Is this action of creatures, one upon the other, subjected also to the action of the divine government?*

The Godness of all matter and energy means that any interaction between forms of Godness effects the Godness of the Cosmos.

- *What is meant by this?*

Actions of all creatures are acts of God and perfect the Godness of the Cosmos and the Godness of each form of God in the Cosmos except those acts intended to prevent another form of God from perfecting their own Godness.

- *Why are creatures thereby more perfect?*

All forms of God are perfect but only those forms of God, such as Human Beings, who are aware of their Godness, may further perfect their Godness and the Godness of the Cosmos by striving to be the best form of Godness unique to themselves and perform acts which help other forms of Godness perfect their own Godness and the Godness of the Cosmos.

The existence of any form of God perfects the Cosmos.

- *When creatures then act one upon the other they are simply executing the orders of God?*

Each form of God is subject to the natural laws of science. Those forms of God which are aware of their Godness are solely responsible for their own actions. No form of God is subject to the orders, plans or providence of any other form of God except the form of God which is consistent with the natural laws of science.

- *Can creatures, in their action one upon the other, be the cause of any particular evil?*

The perfection of the Cosmos must include evil because evil is a form of God. The acts of evil are performed by forms of God and do not detract from the Godness of any form of God or the Godness of Cosmos but rather are necessary for perfecting the Godness of the Cosmos. Without acts of evil the perfection of Cosmos is impossible because perfection is the inclusion of everything that exists in the Cosmos which in its totality is God. Those intentional acts which interfere with the ability for human forms of God to perfect their own Godness and the Godness of the Cosmos are evil.

- *Can any such particular evil happen contrary to the order of divine government?*

Evil is the failure to strive to perfect one's Godness, or inhibit any other form of God of perfecting their Godness. One's Godness is defined in the acts one performs to perfect their own Godness and the Godness of the Cosmos.

- *What is meant by this?*

Human beings are the only form of God that have the responsibility to act to perfect the Godness of the Cosmos. All other forms of God, to our limited knowledge, do not have the capacity to recognize either their own Godness or the Godness of others and therefore do not have the same responsibility to perfect their own Godness. Those forms of God are already perfected and do not have the capacity to act to perfect their Godness or the Godness of the Cosmos. Human beings that have never developed or have lost the capacity to recognize their Godness no longer are responsible to perfect their Godness.

- *Can we, in this life, come to understand this wonderful ordering of divine government in the world?*

The Cosmos is subject to no separate entity's divinity since God is the single organism we call the Cosmos. Once we understand that all things that exist in the Cosmos are a form of God and accept their Godness and understand the natural laws are also a form of Godness we can understand that the perfection of our Godness is essential to the perfection of the Godness of the Cosmos.

- *Where shall we come to see in all its splendour the beauty and the harmony of God's government of the world?*

The perfection of the Cosmos is an infinite process and we can only view it through our own acts that strive to perfect the Godness of the Cosmos.

XXVI. OF THE ACTION OF THE MATERIAL WORLD OR OF THE WHOLE OF THE COSMOS

- *What other beings are there which concur in government?*

Any forms of God, which exist in the Cosmos, and are capable of recognizing their own Godness and the Godness of all that exists in the Cosmos, have the responsibility and capacity to strive to perfect the Godness of the Cosmos.

- *Is then the whole course of nature thus in the hands God for the ruling of the world?*

Natural laws are a form of God or the result from a form of God and have always existed

- *It is then for the realization of the plans of God for the help thereof, that every day the sun rises, day follows night, that the seasons come and go, this in such order that nothing ever disturbs the coming and the going of the days, the months, the years, and centuries?*

The natural laws are defined and not subject to influences from a separate consciousness.

- *Man then is the creature for whom God in some wise has arranged that all other creatures should be subservient to his needs?*

No form of God is subservient to any other form of God. Each form of God serves to perfect the Godness of the Cosmos. Only Human beings, on the Earth, have the capacity to recognize their own Godness and the Godness of all that exists in the Cosmos and the ability to understand how those forms of Godness can be served to perfect the Godness of the Cosmos and the responsibility to perfect their own Godness and the Godness of the Cosmos.

- *Why has God thus acted towards man?*

Knowledge demands responsibility. Human's knowledge of their own Godness makes them responsible for striving to perfect the Godness of the Cosmos.

XVII. OF THE ACTION OF MAN HIMSELF

- *Can man, weak though he be, also help in the action of God towards the government of the world?*

On Earth, only humans have the capacity to perfect their own Godness and the Godness of the Cosmos.

- *How can man thus concur with the action of God in government of the world?*

Humans can strive to perfect their own Godness and therefore, by perfecting their own Godness they can perfect the Godness of the Cosmos.

- *In what way can man co-operate himself for the good of man?*

Once a human recognizes the Godness of all that exists in the Cosmos he is capable of, and responsible for, striving to perfect his own Godness and the Godness of the Cosmos.

- *And how does man serve as an instrument in God's hands as regards the body of man?*

Those human beings that are incapable of recognizing their own Godness either because they have yet to develop that capacity or have by circumstance, lost that capacity do not have the responsibility to strive to perfect their Godness. Perfecting one's Godness can only be done within the unique physical and intellectual limits of one's form of God.

The Nature Of The Omnideistic God

Omnideism is not a matter of faith but a matter of acceptance. Unless one accepts God as a fact Omnideism is irrelevant. Omnideism rejects faith as a condition of Omnideism. Faith determines facts are irrelevant. Time after time facts that contradict beliefs are dismissed in Islam, Christianity, Judaism, Animism, and spiritual religions like Hinduism and Zoroasterism. On its face it would seem that people of faith should be anxious to accept the fact of God in place of the belief.

The spirituality of Omnideism exists only within each form of God that is capable of recognizing their own Godness and the Godness of the Cosmos.

By definition the Omnideistic God is all encompassing. God is judgmental only in the form of the culture through which God temporally exists. Jesus and Muhammed and Ghandi and Abraham were more aware of their Godness and recognized the Godness in all others and may serve or be put forth as examples and teachers of how the human form of God should perform on the form of God we call Earth. The relevant nature of God is the nature of the immediate social environment in which any other form of God exists. The nature of God can only be reflected in the temporal and environmental existence of the form of God as it exists culturally in human form. The nature of God on a planet in a galaxy 100,000,000 light years away from Earth is more likely, than not, to be unrecognizable to the nature of God familiar in Cincinnati. The nature of God is temporal, flexible, relative, and only consistent with the best interests of the forms of God existing in a particular place and time because the nature of God, as the single organism known as the Cosmos is not perceivable or comprehensible to any individual form of God Who exists as the aggregate of the Cosmos.

The Omnideistic God obliges every human being to be entirely responsible for their own actions. As a particular and unique form of God each human being has the ability to perfect their own life in the knowledge of their own Godness and the acceptance and recognition of Godness in everything that exists in the single organism which is the Cosmos.

65283651R00071

Made in the USA
Middletown, DE
24 February 2018